Developmental Disabilities:
A Training Guide

AMERICAN HEALTH CARE ASSOCIATION/CBI PUBLISHING COMPANY
SERIES IN HEALTH CARE MANAGEMENT

RESIDENT CARE MANAGEMENT SYSTEMS
Erwin Rausch
Maria Menna Perper

BASIC ACCOUNTING AND BUDGETING FOR LONG TERM CARE
FACILITIES
Jerry L. Rhoads

THE SOCIAL WORKER IN THE LONG-TERM CARE FACILITY
Shirley Conger
Kay Moore

DEVELOPMENTAL DISABILITIES: A TRAINING GUIDE
Henri Deutsch
Sheldon Bustow

Developmental Disabilities: A Training Guide

Henri Deutsch, Ph.D.
and
Sheldon Bustow
Carol Winters Wish, Ph.D.
Joel Wish, Ph.D.

CBI Publishing Company, Inc.
51 Sleeper Street
Boston, Massachusetts 02210

Production Editor: Donna L. Miner
Interior Designer: Jack Schwartz
Cover Designer: Charles G. Mitchell
Compositor: Publications, Inc.

Library of Congress Cataloging in Publication Data

Deutsch, Henry, 1949–
   Developmental disabilities.

   Bibliography: p.
   Includes index.
      1. Mentally handicapped children—Education
—Handbooks, manuals, etc. 2. Physically handi-
capped children—Education—Handbooks, manuals,
etc. I. Bustow, Sheldon M., 1948–   II. Title.
LC4601.D458   371.91   81-10164
ISBN 0-8436-0851-X   AACR2

Printed in the United States of America
Printing (*last digit*): 9 8 7 6 5 4 3 2 1

To Marilyn and Jonathan

Contents

Acknowledgments

This book was the brain-child of Gloria Morrow and Sheldon Bustow, without whose assistance this book would not have been written. Special thanks to Gloria Morrow for her editorial role. In addition, I would like to thank Chris Sanders and Marilyn Deutsch for their critical review of the chapters. Also, a special acknowledgment is due to James Bailey and Peter Damok for their photographic assistance. Finally, recognition must be made of the fine typing and spelling skills of Ann McPherson and Mary Bass.

Foreword

The 1980s will be a decade of marked change for developmentally disabled citizens and for those who work in this challenging field. Large, antiquated, dehumanizing institutions are shrinking, and many will cease to exist in the future. Retarded and other developmentally disabled persons are seldom sent to state institutions today. Yet their need for long-term living alternatives has never been greater. Parents' groups, advocacy organizations, court rulings, and state and federal statutes are leading us away from institutional care, believing that proper services, which are both humane and growth-promoting, can only be rendered in smaller, more personalized community settings.

Two things remain unchanged. First, client needs persist. Second, the need for primary care givers (direct care staff) will continue as long as client needs exist. In other words, settings such as institutions may change, but the need for committed and caring staff to operate alternative living settings can only grow.

The authors of this book offer a unique perspective regarding the value of direct care personnel. This is not one of those demeaning "skills only" pamphlets that have proliferated the field. Instead, Dr. Deutsch and Mr. Bustow understand the essential worth and dignity which should be found in direct care staff, and offer a readable and comprehensive overview of developmental disabilities for these personnel. This book, if properly utilized, could comprise the core of a comprehensive and ongoing in-service training curriculum for primary care givers working in the various living alternatives for our developmentally disabled citizens.

The book represents a meaningful contribution to the field, which emanates from the unusual combination of hands-on clinical and administrative theory and experience offered by the authors.

Administrators, clinicians, and direct care staff can all benefit from reading this work. It provides a solid base for improving the everyday lives of developmentally disabled persons through the everyday work of caring and committed staff.

Robert L. Carl, Jr., Ph.D.
Associate Director for Retardation
Rhode Island Department of
Mental Health, Retardation,
and Hospitals

Preface

This book is designed as a training manual for direct care staff working with developmentally disabled individuals. The information found in this book can be helpful to staff in a variety of residential settings, including large public institutions, skilled nursing facilities, intermediate care facilities for the mentally retarded, private residential facilities, and group homes. Thus, regardless of the titles that direct care staff have in these settings (nurse's aides, developmentalists, trainers, child care workers), this information can be valuable in working with clients in any residential setting.

In writing this book we have attempted to include the current trends in the field of developmental disabilities, placing a major emphasis on training and client programming. In addition, one chapter deals with the stresses often encountered by families with a developmentally disabled child. It is hoped that this chapter will give staff a fuller understanding of the dynamics that occur within the family, and so allow them more sensitivity when working with their clients' families.

The last chapter of the book is about "staff burnout." Although administrators frequently discuss this topic, vital information often does not filter down to the direct care staff. Since the field of developmental disabilities is marked by a high degree of staff burnout, it is imperative that the staff are familiar with this concept and the methods available for avoiding becoming "burned out."

Each chapter contains a group of questions relating to the material discussed. Students' familiarity with the concepts found in each chapter can be enhanced by utilizing these questions, as well as the "Activities for Thought and Discussion."

These activities can often be done in a group setting, allowing students to use the chapter's ideas actively, rather than merely reading them. The instructor may also quiz students on selected items from these questions and activities and thus gauge the degree to which they have incorporated the ideas.

Henri Deutsch
Worthington, Ohio
July, 1980

Introduction to Developmental Disabilities

1

Sheldon Bustow

This manual has been written to teach you the basic information necessary to provide the care, treatment, and programs needed by the developmentally disabled clients under your charge. While the term *developmentally disabled* may be new for you, the clients and their needs are not. Mental retardation, epilepsy, cerebral palsy, autism, and related conditions are now grouped together as developmental disabilities.

This manual contains information that you will find helpful in working with your clients on a day-to-day basis. This first chapter explains the conditions called developmental disabilities: what causes them and how they affect a person. (You will discover that people with the same problem can be very different.)

A developmental disability is not a new disease, but rather, a new classification of several conditions known and treated for some time now. The three major conditions chosen for discussion in this manual are mental retardation, cerebral palsy, and epilepsy. These are the conditions you will most likely be involved with on a daily basis. Note that we refer to *conditions* instead of *diseases.* This is an important distinction, because you will be helping the person and not curing the disability.

DEVELOPMENTAL DISABILITIES CRITERIA

Over the past ten years Congress has passed a number of laws to provide a common plan of action for all the states in meeting the needs of individuals with developmental disabilities. The most recent definition of developmental disabilities is as follows:

> A developmental disability is a severe, chronic disability of a person which:
> 1. is attributable to a mental or physical impairment or combination of mental and physical impairments;
> 2. is manifest before age 22;
> 3. is likely to continue indefinitely;
> 4. results in substantial functional limitations in three or more of the following areas of major life activity:
>   a) self-care
>   b) receptive and expressive language
>   c) learning
>   d) mobility
>   e) self-direction
>   f) capacity for independent living, or
>   g) economic self-sufficiency; and
> 5. reflects the need for a combination and sequence of special, interdisciplinary or generic care, treatment or other services which are:
>   a) of lifelong or extended duration and
>   b) individually planned and coordinated. (Public Law 95–602, 1978)

A condition must meet all five of Congress's criteria to be classified as a developmental disability.

Caused by Mental or Physical Impairment   The condition must be the result of a mental or physical impairment or a combination of the two. A child whose parents are having marital problems may have some difficulty in school as a result of the disruptions at home. This, however, would not be considered a developmental disability, since neither mental nor physical impairments caused the school problems. On the other hand, if mental retardation were the cause of the child's failure in school, then he/she would be considered developmentally disabled.

Must Occur before Age Twenty-Two   The condition must show itself before one reaches twenty-two years of age (except for mental retardation, which must occur within the first eighteen years of life). A middle-aged man who has severe brain damage from an automobile accident and experiences seizures would have a form of epilepsy, but it would not be developmental in nature, since it occurred after age twenty-two. If a six-year-old suffered brain damage from a car accident, however, and the resulting seizures continued into adulthood, then this person would be developmentally disabled, since the physical damage first occurred before his or her twenty-second birthday.

Is Likely To Continue Indefinitely The condition must be continuing in nature and expected to be a disability for some time. A broken arm may hinder activity for a while, but it should heal so as not to be a burden or disability for life.

Limits Daily Functioning The condition must result in the limitation of at least three major activities of life that the rest of us take for granted. These areas include self-care (ability to feed or bathe oneself, for example), effective communication, ability to learn, freedom of movement, and economic self-sufficiency. For example, an individual who is visually impaired (blind) probably would not be developmentally disabled, because the inability to see would not necessarily create difficulties in learning and independent functioning. Although a blind individual may encounter difficulties in daily living, many of them can be resolved with training. Thus, if a visual impairment does not limit functioning in at least three daily life areas, then it is not a developmental disability.

Needs Special Care from a Number of Professionals Finally, a person with a developmental disability requires care by an interdisciplinary team of professionals (see chapter 3 for more details on a *team approach*). These professional services provide assistance and support for an extended period of time, and they must be planned to meet the individual's needs. A person who needs only the emotional support of the family to be able to attend school and then proceed to employment would not have a true developmental disability.

CAUSES OF DEVELOPMENTAL DISABILITIES

Developmental disabilities are caused by many different diseases, accidents, and genetic defects. For many individuals, the specific cause is unknown. Although these primary causes of developmental disabilities usually cannot be corrected, the condition can be lessened in its severity so that the individual can pursue a more normal lifestyle.

The primary causes of developmental disabilities can be grouped into three categories: *prenatal, perinatal,* and *postnatal.* We will discuss each category and include some examples that may be helpful to you in understanding your work.

Prenatal Causes

Prenatal means "before birth." Thus, prenatal causes of developmental disabilities are those that occur during pregnancy and affect the development of the fetus (unborn child). Prenatal harm can be caused by genetic factors and by physical damage (Wyne and O'Connor 1979).

Genetic Factors   One type of prenatal cause is *genetic or hereditary* in nature. This means that the basic chromosome code of the fertilized egg contains some abnormality that leads to a flawed development of the fetus or unborn child. Chromosomes are the parts of our cells that dictate basic characteristics, such as hair color and height. Genetic factors were once thought to be the major reason for developmental disabilities, but now it is known that they are responsible for only a small number of cases. Perhaps the most commonly known genetic cause is Down's syndrome or mongolism. This syndrome is caused by the presence of an extra chromosome in the fertilized egg; where a normal person has forty-six chromosomes, the person with Down's syndrome has forty-seven. Factors such as the mother's age at conception and a history of the defect in the mother's family contribute to a child's increased possibility of having Down's syndrome. Mental retardation, more often than epilepsy and cerebral palsy, is caused by genetic factors.

Physical Damage   Prenatal causes may also be due to diseases or accidents that occur during pregnancy. In these cases the fetus begins a normal development but is damaged or altered by an infection or an accident. Some of the most commonly known causes are rubella (German measles) infection, alcohol usage by the mother, drugs that may affect the fetus (heroin and thalidomide are two examples), and physical abuse of the mother. Most of the prenatal causes are preventable if the mother follows proper precautions. These causes are most damaging in the first three months of pregnancy, when the fetus is developing most rapidly and is most vulnerable. Research indicates that prenatal damage is the second largest cause of all developmental disabilities and may well be the largest cause of mental retardation.

Perinatal Causes

Perinatal causes are those that occur during the process of birth itself. Perinatal factors appear to be the main causes of cerebral palsy and epilepsy and a leading cause of mental retardation. Most common in this group are anoxia (where the infant's brain is damaged by not receiving enough oxygen), physical harm to the brain by the use of forceps in delivery, and damage caused by a narrow birth canal (Wyne and O'Connor 1979). The decreased use of forceps, the increased use of cesarean section delivery techniques (to avoid the compression of the brain in a narrow canal), and the monitoring of the fetus during labor (to spot problems in blood and oxygen supply) have led to hopes for decreasing these damaging traumas.

Postnatal Causes

Postnatal causes are those that occur after birth at any time during the developmental period of life. Such well-known factors as lead poisoning, child abuse (either physical abuse or neglect), and childhood meningitis (an infection of the central nervous system) can damage the brain to varying degrees of severity. Accidents (e.g., car, sports) can also cause brain damage. Although postnatal causes account for a rather small percentage of the total developmental disability cases, they are most alarming, because in most situations the damage is preventable.

MENTAL RETARDATION

The currently accepted definition of mental retardation, the largest group of developmental disabilities, is as follows:

> Mental Retardation refers to significantly subaverage general intellectual functioning existing concurrently with deficits in adaptive behavior, and manifested during the developmental period. (Grossman 1973)

Mental retardation, then, refers to a level of actual functioning and not a specific disease of cause. When you work with a client, it is not usually important to know the original cause of the condition. What matter more are the potential of the individual and the obstacles to be overcome in realizing that potential.

Characteristics of Mental Retardation

There are three major parts of this definition of retardation that must be understood: *general functioning, adaptive behavior,* and *developmental period.*

General Intellectual Functioning General intellectual functioning is measured by an intelligence test (which provides an intelligence quotient, or IQ) or other measure of level of development. The results of this test must be significantly below average before someone can be labeled mentally retarded. (A complete listing of IQ levels is provided in Exhibit 1.1.) It is important to understand that general intellectual functioning alone does not prove or indicate mental retardation, since intelligence is only one part of the definition.

Adaptive Behavior Adaptive behavior indicates the ability of an individual to function as others within the same age and cultural group. Because expectations of behavior vary by age and cultural background, behavior that is appropriate for one individual often would not be appropriate for another.

To indicate mental retardation, problems in adaptive behavior must be serious and must coincide with a significantly low IQ score. A deficit in

Level of Retardation	Measured IQ	
	Stanford-Binet	**Wechsler**
Not retarded	68 and above	70 and above
Mild	67–52	69–55
Moderate	51–36	54–40
Severe	35–20	39–25 (extrapolated)
Profound	19 and below	24 and below (extrapolated)

Exhibit 1.1. IQ levels of mental retardation. (Stanford-Binet and Wechsler are two commonly used intelligence tests.)

intelligence alone or in adaptive behavior alone does not indicate that someone is mentally retarded; both must be present (see Exhibit 1.2.). Thus, people with average or above average intelligence who do not have good adaptive behavior should not be labeled retarded. They may be emotionally disturbed or senile or absent-minded professors. On the other hand, people who score quite low on an intelligence test may have good adaptive skills (they work, take care of themselves, and stay out of trouble). These people should not be labeled retarded either.

Adaptive behavior is measured by an adaptive behavior scale. There are several different kinds of scales available, but all allow us to compare an individual's level of functioning with that of others, so that we can see the degree of adaptive impairment.

Developmental Period The first eighteen years of life are known as the *developmental period*. (The definition of developmental period when mental retardation is concerned is slightly different than the developmental period described earlier for developmental disabilities.) Someone whose condition is caused by an accident or disease after age eighteen could not correctly be called mentally retarded, but may show the same needs and disabilities as someone who is. You may at times, in a long-term care setting, be caring for just such an individual, and the treatment and programs would be similar.

Levels of Mental Retardation

There are four levels of mental retardation as measured by intellectual level and adaptive behavior. Exhibit 1.1 outlines the degree of intellectual functioning for each of the four levels. You should note that even with the same measured IQ and the same adaptive behavior, individuals will often show quite different

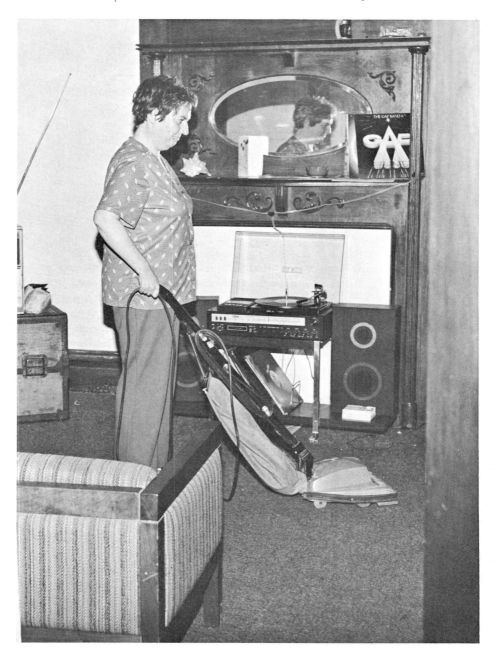

Figure 1.1. Learning to care for one's home improves adaptive behavior.

		Adaptive Behavior	
		Low	High
Measured Intelligence	Low (below IQ of 70)	Mentally retarded	Not mentally retarded
	High (above IQ of 70)	Not mentally retarded	Not mentally retarded

Exhibit 1.2. Interaction of measured intelligence and adaptive behavior in determining mental retardation.

behavior and skills. Much of the difference is due to past experiences, environment, emotional differences, and personality differences. It is important to avoid putting people into stereotyped categories; you should expect and respect individual differences. Often individuals may be labeled as functioning at one level, but behave as if they were capable of more or less. These variances point to the need to plan programs for each individual instead of fitting people into one group because of prior labels that have been imposed. Each person has his or her own areas of strength and weakness.

Profound Mental Retardation Profound mental retardation is the lowest functional level. Profoundly retarded individuals need constant attention and help with their daily life skills. They reside in state institutions, in nursing homes, or with their parents. Many communities have developed day and home training programs that allow these people to remain in their homes or in small residential settings in their own communities, rather than in institutions. Some profoundly retarded individuals have complicated medical problems that require medical attention beyond that which can be provided in the home. These people are often best served in facilities that provide nursing care.

Profoundly retarded individuals with medical problems in addition to their mental retardation often lose muscle control and thus cannot control body movements. Frequently these difficulties stem from a lack of early-infant and preschool stimulation. They often result in deformities and make later therapy much more difficult. Therefore, it is important for the resident to be moved frequently in a variety of positions, exercised, and stimulated. A small investment of your time in such actions is much better, for you and the client, than dealing with bedsores, stiff muscles, and clients who have great difficulty helping themselves in later life.

During the school age years, active and consistent training can assist the profoundly retarded in developing improved self-care skills, such as feeding, dressing, bathing, and toilet use. In some cases language training, social skills, and personal controls can also be emphasized. Do not assume that the profoundly retarded will not learn. All people are capable of learning to some degree. Of course, helping profoundly mentally retarded individuals improve their skills is a slow process and will take much time and effort on your part. The rewards for you and these clients, however, can be great.

When you are dealing with profoundly retarded adult residents, you are working with the results of their prior training or lack of training. If they cannot do some things for themselves, it is not too late to start training, even though a client's physical problems may seem overwhelming. The coordination between yourself, the physical therapist, the nurse, and other professionals becomes very important in following a total, comprehensive program designed for the individual. The physical therapist, for example, may show you some easy exercises to relax constricted muscles. Stimulating the clients can be accomplished by having them touch different kinds of surfaces, providing different images for them to look at, and communicating through any means possible. Talking to clients about what is going on around them, even if they don't seem to understand, is far better than treating clients as if they did not exist. Remember, if you had been between two sheets, staring at a white ceiling and alone with your own thoughts for years, then you too would probably be very receptive to new experiences.

For the profoundly retarded adult who has been fortunate enough to receive and respond to early stimulation, you can build upon that progress in many ways. Important activities include working on communication skills, allowing the client to choose between leisure activities, and sometimes even teaching simple work skills.

On the whole, the profoundly mentally retarded are not difficult to work with on a daily basis. You must recognize the potential of these individuals, and indeed, the fact that they *are* individuals, to appreciate their development and self-care potential.

Severe Mental Retardation Severely retarded individuals live in a variety of settings, ranging from their own homes and group homes to nursing facilities and institutions. The primary goal of their programs should be to develop the basic self-help, social, and simple work skills that will enable them to live in supervised arrangements outside of institutional facilities.

When these individuals are infants, their programs will closely resemble those for the profoundly retarded. You will spend a lot of time on floor mats, repositioning clients and stimulating them with various materials to develop not only their senses, but also their curiosity. For example, you should take the time not merely to feed clients but to teach them how to feed themselves.

Other self-help skills like bathing, dressing, and toileting will also be worked on. Do not expect, however, that this will lead to overnight success. Progress in all these areas occurs slowly, but it does occur.

When the client reaches school age, efforts to teach self-help skills become primary. Also, you will begin to teach social skills, such as playing in groups and speech/communication skills. This will be an important time for the client and for you. With mastery of these skills, the client will need much less of your physical energy and will be able to communicate desires and wishes with greater ease. Once these basic skills are learned, other skills can be taught to provide more independent behavior and to foster feelings of self-worth.

By adulthood the client should be ready to learn simple work skills and the remaining social skills necessary for community living and sheltered employment. If the client is still in an institutional setting at this age, the team will have to assess how he or she functions. The team should then develop a special program to teach the self-help skills (toileting, dressing, bathing), if these were not learned earlier. It is never too late to start programming with clients; just start at the right place.

You will find many rewards in working with severely retarded clients, not the least of which is sharing their joy in the mastery of skills we take for granted. You will find that coming to work every day is much more pleasant when you are involved in teaching people new skills and new ways to enjoy life and themselves.

Moderate Mental Retardation For most persons with a condition of moderate mental retardation, your efforts will be less physical, but focused more on helping them act like people their age. By school age, most moderately retarded individuals are capable of achieving the basic personal self-help skills and will be ready for the kinds of programs experienced by the normal school-aged population.

With school-aged children, you work to teach grooming, preemployment training, communication skills, social adjustment, and those kinds of behavior necessary for survival in our society. Many of your clients may be with you only in the evenings and weekends, as they attend school or workshops. Two of your more important objectives during this time are to provide as normal and home-like an environment as possible and to coordinate your efforts with those of the school or workshop staff.

By early adulthood many moderately retarded individuals will be living in community residences of varying levels of supervision and be employed in sheltered workshops or competitive employment situations. Adults who as children did not have the opportunity to learn the skills necessary to function in society still have the potential for learning these skills. Your job is to assist them in doing as much as they are able to.

Figure 1.2. Many moderately and mildly retarded people live in the community and enjoy their leisure time.

Mild Mental Retardation Mildly retarded adults are generally capable of living in the community in group homes, supervised apartments, or on their own. Generally, these people would not live in a long-term health care facility unless they had a health problem.

During their school years, the mildly retarded usually attend special education classes in the public schools. Here they are introduced to the basic academic skills (arithmetic, reading, social studies) and to such daily life skills as budgeting money, shopping, and home economics. As these children go through high school, vocational training is often introduced.

In many school districts the mildly mentally retarded are *mainstreamed* into regular classes. This means that instead of being placed in a special education class, these children attend the same classes as everyone else. This can vary, depending on an individual's needs, from one or two classes (gym, music, etc.) to the entire day. Supporters of mainstreaming believe that the mentally retarded should not be segregated into separate classrooms because they can learn as much in the regular classes as they can in the special education classes, while they are able to socialize with other children to a greater degree.

Mildly retarded adults can generally be employed and be self-supporting. They can often hold a wide variety of jobs, and can thus contribute in a

11

meaningful way to society. Many mildly retarded men and women get married and have families as well (see chapter 10 for more details on sex and marriage).

EPILEPSY

Epilepsy, the second group of developmental disabilities described in this book, is a "group of effects resulting from a disturbance in the brain's electrical activity" (Silverstein and Silverstein 1975). Electricity in the brain? Yes, messages in the brain are sent by a form of electricity. For most people, the flow of electrical discharges from one nerve cell to the next is controlled by a complex system of checks and balances. When these controls break down, however, an uncontrolled discharge produces a seizure. An individual's seizure activity can vary from day to day because of illness, fatigue, or emotional stress.

Epilepsy can occur in conjunction with several other developmental disabilities, such as mental retardation or cerebral palsy, or it may occur on its own. When an individual has epilepsy as well as another developmental disability, the same type of brain damage has usually caused both disabilities. Epilepsy does not "cause" mental retardation or cerebral palsy. In fact, the large majority of the more than one million people with epilepsy are not mentally retarded (Wyne and O'Connor 1979).

Types of Seizures

Epilepsy may involve different types of seizures. Each individual will usually have a specific type and only occasionally a combination of types. The basic types of seizures are *grand mal, petit mal, psychomotor,* and *akinetic.*

Grand Mal Seizures   A grand mal seizure is the most common type and is the classic form: an individual suddenly loses consciousness and topples over, becoming very stiff. This first phase is called the *tonic phase.* Once the stiffness passes, the second phase begins: the body starts jerking motions, which may lead to biting one's tongue, bumping into close objects, or seemingly pushing away those there to help. This is called the *clonic phase.* During this phase, the person is still completely unconscious. The body soon goes limp and the seizure is over. Such a seizure can last from thirty seconds to over thirty minutes. After the seizure the person usually is very tired and should be placed in bed or on a sofa to rest. Headaches, soreness, and restlessness are some of the aftereffects. Although people generally recognize that they have had a seizure, they usually cannot remember all the details of the event.

Grand mal seizures are sometimes preceded by an *aura,* a specific sensation that warns the individual that a seizure is about to occur. The aura may be a specific smell, the tingling of a certain limb, a buzzing sound, or other such sensations. If a person can learn to recognize such sensations, or if you can, then the seizure can be anticipated and the person can lie down or otherwise prepare for it. The recognition of an aura does not allow for any preventive measure beyond this. If you have a client with grand mal seizures, the guidelines in Exhibit 1.3 might prove helpful.

First Aid for Epileptic Seizures

A major epileptic seizure is often frightening to watch but usually lasts only a few minutes. It does not require expert care, and seldom is anything gained by taking a person to a hospital emergency room. The following procedures should be followed:

1. Keep calm. Once a seizure has started, it cannot be stopped. The seizure will run its course. Remember, the individual is not in pain.

2. If you can, ease the person to the floor and try to prevent him from striking his head or body against any hard, sharp, or hot objects. It is important that you do not interfere with his movements. Remember, you do not need to restrain him physically.

3. When the person becomes quiet, turn him on his side and point his face downward so that saliva or vomitus can drain out and is less likely to be inhaled.

4. Do not place anything between the person's teeth. There may be violent teeth clenching as part of the seizure. Teeth may be broken or gums injured if you attempt to put objects into the mouth.

5. Do not be frightened if the person having a seizure seems to stops breathing momentarily. Breathing will be resumed spontaneously.

6. After the seizure stops and the person is relaxed, he should be allowed to sleep or rest if he wishes. He usually returns to his normal activities as soon as he feels capable of doing so.

7. If the jerking of the body does not stop within five minutes or keeps recurring, medical assistance should be obtained.

Exhibit 1.3. Procedures for treating grand mal seizures (based on those recommended by the Epilepsy Association of Franklin County, Ohio [1977]).

Petit Mal Seizures The second type of epilepsy involves petit mal seizures. These are brief (five to ten seconds) but frequent (two to one hundred times a day) attacks of impaired consciousness (Wyne and O'Connor 1979). This type of seizure is relatively mild, and the individual and those nearby may not even be aware of its existence. Usually the person will not fall down, but will appear to stare into space or appear to be daydreaming. When the seizure is over, the individual can go on with normal activity as though nothing had happened. Petit mal seizures usually affect children and either disappear with age (around the teen years) or turn into grand mal seizures in adolescence and adulthood.

Psychomotor Seizures   There are two different types of psychomotor seizures. The first (mid-temporal lobe seizures) usually occur in children. These may involve the child's reporting unusual sights or sounds, having a racing pulse, perspiring, salivating, or having unusual tongue movements. In addition, the child may engage in behaviors that appear to be out of place (e.g., buttoning and unbuttoning a shirt).

Adult psychomotor seizures (anterior temporal lobe seizures) may take a variety of forms. The adult may hear or see things that are not physically present or feel fear or rage for no apparent reason. If the seizures continue, the person may say or do things that are out of place (e.g., open an umbrella inside a house) and that cannot later be recalled (Wyne and O'Connor 1979).

Akinetic Seizures   Akinetic seizures are sometimes called *drop seizures* because the person falls forward or backward without warning. Obviously, head injuries are possible unless people with these seizures wear some form of head protection or receive treatment. These seizures are more common among children than adults and may occur up to several hundred times a day (Wyne and O'Connor 1979).

Treatment

There are a number of drugs that can help control seizures, although doctors are not entirely sure why these drugs, called *anticonvulsants,* work. For grand mal seizures phenobarbital, Dilantin, Depokane, and Mysoline are commonly used, while Diamox, Zarontin, and Tridione are used for petit mal seizures. Since persons react very differently to these medications, the physician may have to prescribe different drugs for each individual. Thus you may have three patients with grand mal seizures, each taking a different drug. Drugs should never be substituted or stopped except under the direct order of a physician.

The most important thing to remember is that epilepsy in and of itself is not a truly disabling condition. You will rarely see a patient in a residential setting who is only epileptic; he or she will usually have epilepsy in conjunction with other disabilities, such a mental retardation or cerebral palsy. If you treat epilepsy as a condition that need not hamper one's life, then you will go a long way in assisting the individual in realizing this as well.

CEREBRAL PALSY

The third group of developmental disabilities to be discussed is cerebral palsy, a group of conditions that have in common a disorder of movement or posture resulting from damage to the brain or the central nervous system. In other words, damage to the brain causes some loss of control over muscles in one or more parts of the body. Thus, while a person with cerebral palsy may have normal leg, arm, or speech muscles at birth, abnormal messages sent by the brain can lead to spastic movement, constant or uncontrollable twitching, and impaired speech. This brain damage can occur before, during, or after birth.

As with mental retardation and epilepsy, the degree of disability varies greatly among individuals. A person with cerebral palsy may have minor, almost unnoticeable disabilities or may need multiple appliances or assisting equipment to sit in a chair or move about. It is estimated that some 700,000 Americans have varying degrees of this disability (United Cerebral Palsy Association, 1979).

Do not assume that simply because your patient has cerebral palsy, he or she is also mentally retarded or epileptic. Many potentially productive and talented people have been stifled in the past by this kind of hasty and erroneous assumption.

Types of Cerebral Palsy

There are three basic types of cerebral palsy. The first type, *spastic,* refers to those conditions where the individual moves with great difficulty and stiffness. Many of the younger patients you work with may move like older people with arthritis. This is an example of spasticity.

The second type of cerebral palsy, *athetoid,* is present when the individual moves uncontrollably and involuntarily. Often this leads to an almost rhythmic motion of a limb that seems to go on endlessly and may be particularly frustrating to you when you are attempting to help the patient. You must remember, however, that this motion is involuntary and not an effort to annoy or disturb you. The patient is probably as frustrated as you in trying to control it.

The third type, *ataxic,* is shown in a basic incoordination of movement, lack of balance, and poor depth perception (ability to perceive space and distance). These patients may often fall for apparently no reason and have difficulty in feeding themselves, since they lack the depth perception we take for granted.

Management is a better word than *treatment* for the services you will provide to these patients. The goal is not to treat a basic disease or disorder, but rather to help the patients manage or cope with their limitations. The management program for motor difficulties will consist of three steps: relaxation of the muscles, training of the muscles for voluntary movement, and channeling of these voluntary movements into appropriate activities.

The physical therapist will design a program of exercise and assist you in learning the specific technique for each patient. It is important to help the patient perform these exercises in consistent, smooth, and regular patterns to achieve maximum benefit. Do not expect overnight success or miracles, but relatively soon you will begin to find the patient more responsive to your directions and eventually more self-sufficient.

QUESTIONS

1-1. What is the difference between mental retardation and mental illness?

1-2. Can someone have more than one developmental disability?

1-3. What are some a) prenatal, b) postnatal, c) perinatal causes of developmental disability?

1-4. What would you do if one of your clients had a grand mal seizure?

1-5. Name and describe the three types of cerebral palsy.

1-6. Name and describe the different types of epileptic seizures. What is an aura?

Principles of Care and Treatment

2

Mentally retarded and other developmentally disabled persons have been regarded as deviants in our society, as not fitting into accepted social and work roles. As a result these people have often been separated, in one way or another, from the rest of the community. One of the effects of this segregation is to continue the myth that the developmentally disabled are deviants. Many people have the notion that the mentally retarded are different from others, meaning that they are less than human. When this happens other *false assumptions* arise about the developmentally disabled: that they have different emotional needs than the rest of us, that it does not matter if they are treated in less than human ways, that they do not benefit from people's caring about them or from other normal experiences, that they cannot grow or develop or learn. The next few chapters will reveal not only how wrong these assumptions are, but also how harmful they can be to the people you were hired to help.

PEOPLE WORKING WITH PEOPLE

A basic premise of this book is that you will be working with people—people with emotional needs, people with physical needs, people who laugh, people who cry, people who can enjoy life, themselves, and others, people who can grow and learn and develop, people who are affected by what goes on around them, people who have a place in the world. If you keep these ideas in mind, you will find that your job will be extremely rewarding for you, as well as for the people with whom you are working.

CARE AND TREATMENT OF THE MENTALLY RETARDED: A BRIEF HISTORY

A review of the historical trends in the care and treatment of the mentally retarded must include some mention of the public institutions. Between 1848 and 1880, professionals in the field believed that the purpose of working with the mentally retarded should be to make them "normal." To accomplish this task, they felt that mentally retarded persons must be separated from society for a short period of time to be rehabilitated, and then be returned to the community. This was the beginning of public residential institutions in the United States.

In 1848 Massachusetts opened the first institution in the country and was quickly followed by Pennsylvania, Ohio, and New York. These early institutions were quite progressive. They considered the medical, educational, and social aspects of mental retardation and attempted to foster independence among their clients (Adams 1971).

Institutions Begin To Change

By 1880 the character of the institutions was changing, from single purpose to multipurpose. In addition to rehabilitation and the return to the community,

Figure 2.1. Many old institutions for the mentally retarded still have large buildings, housing many people under the same roof.

they were beginning to provide long-term placement and custodial care. Through their own farms and workshops the institutions became self-supporting and, as a result, more isolated from the rest of society. This isolation was intended, because professionals now believed that the mentally retarded should be protected from society. It was felt that separating these individuals from the community was the best way to prevent the rest of society from taking unfair advantage of them.

Around the turn of the century, this explanation for the isolated location of the institutions was reversed; now it became "to protect society from this inferior stock." This was the period of the "Eugenic Scare" started by Henry Goddard, who stated that mental retardation was an inherited condition that led to poverty, prostitution, and crime. Goddard believed that the cure for mental retardation, and a method of preventing additional mentally retarded persons from being born, was to segregate the retarded from society or to sterilize them. We now know that there are many causes of mental retardation, and that isolating mentally retarded persons does not prevent the birth of other people with handicapping conditions.

As a result of Goddard's influence, the early part of this century witnessed the beginning of large population growths within the institutions (Roos 1969). Additionally, along with attitudes toward the retarded that deprived them of many basic rights (e.g., freedom of choice, where to live, who to marry, where to work) came the notion that the mentally retarded were less than human. If that were true, then mentally retarded persons could be treated as nonhumans. This was the beginning of a dehumanization process that continues today in some (but certainly not all) institutions.

Dehumanization

Dehumanization can take many forms, some quite obvious and others not quite as obvious. It can include physical environments where many people are housed together with little or no privacy. This may include large wards, communal showers, and toilet facilities without partitions. Or it may involve the physical care of the client, such as not being allowed to have personal possessions or clothing, being forced to have the same haircut as everyone else, not having adequate food or choice of foods, not being able to eat in a comfortable and relaxed manner. Dehumanization may also occur in the way staff interact with residents. This may be the most harmful to the people you are serving and can take such forms as

a) not talking to or belittling clients because it is felt that they cannot understand what is said to them;

b) believing that clients cannot learn or become more independent, and therefore not bothering to train clients how to eat, dress, bathe, or toilet themselves; and

19

    c) believing that clients cannot do anything or enjoy anything, and therefore not providing any stimulation or programming for them.

The New Era

The 1960s opened with a new philosophy for the care and treatment of mentally retarded persons. At first there appeared to be a reaction to the dehumanizing conditions of the institutions. A shocked public became aware of the horrors of institutional life, especially the back wards (Blatt and Kaplan 1966, Wolfensberger 1969), and a movement soon followed on the national level to combat these horrors.

In 1962 President Kennedy appointed a special panel of experts, later to become the President's Committee on Mental Retardation, to make recommendations for legislative reform in the area of mental retardation. In February 1963 the president sent a message to Congress based on this panel's report, which recommended establishing "a national program to combat mental retardation." As a result of President Kennedy's influence, money became available for research in the field of mental retardation, as well as grants to improve conditions within the institutions.

CURRENT TRENDS

After this initial push, interest in the field of mental retardation grew steadily. New philosophies and beliefs arose. It is now believed that the mentally retarded can be habilitated (able to learn new skills and to become more independent) and that the habilitation process does not necessarily have to occur within the institutions. As a result, the number of people in large institutions is decreasing, and community training programs for the mentally retarded are expanding. Whether we work with people in the large institutions or in smaller facilities in the community, we must recognize that our clients have certain undeniable rights, as expressed in Exhibit 2.1.

What this recognition means is that a mentally retarded person should have the same rights as any other citizen of the same country. These rights include the availability of all of the habilitative services (medical, psychological, educational, training) that would allow that individual to become more independent.

The mentally retarded person also has the right to live in the community and to participate in as normal a life as possible (economic security, job, recreation). A mentally retarded individual who needs some assistance in protecting his or her well-being has the right to a guardian.

These rights may be modified to suit the needs of the individual only if the proper legal safeguards are used, and only after the individual is evaluated by experts. If a person's rights are modified or denied, the decision to do so must be reviewed on a regular basis.

Declaration of Rights for the Mentally Retarded

Whereas the universal declaration of human rights adopted by the United Nations proclaims that all of the human family without distinction of any kind, have equal and inalienable rights of human dignity and freedom.

Whereas the declaration of the rights of the child, adopted by the United Nations, proclaims the rights of the physically, mentally, or socially handicapped child to special treatment, education, and care required by his particular condition.

Now, therefore, the International League of Societies for the Mentally Handicapped expresses the general and special rights of the mentally retarded as follows:

Article I: The mentally retarded person has the same basic rights as other citizens of the same country and same age.

Article II: The mentally retarded person has a right to proper medical care and physical restoration and to such education, training, habilitation, and guidance as will enable him to develop his ability and potential to the fullest possible extent, no matter how severe his degree of disability. No mentally handicapped person should be deprived of such services by reason of the costs involved.

Article III: The mentally retarded person has a right to economic security and to a decent standard of living. He has a right to productive work or to other meaningful occupation.

Article IV: The mentally retarded person has a right to live with his own family or with foster parents, to participate in all aspects of community life, and to be provided with appropriate leisure time activities. If care in an institution becomes necessary it should be in surroundings and under circumstances as close to normal living as possible.

Article V: The mentally retarded person has a right to a qualified guardian when this is required to protect his personal well-being and interests. No person rendering direct services to the mentally retarded should also serve as his guardian.

Article VI: The mentally retarded person has a right to protection from exploitation, abuse, and degrading treatment. If accused, he has a right to a fair trial and full recognition being given to his degree of responsibility.

Article VII: Some mentally retarded persons may be unable, due to the severity of their handicap, to exercise for themselves all of these rights in a meaningful way. For others, modification of some or all of the rights is appropriate. The procedure used for modification or denial of rights must contain proper legal safeguards against every form of abuse, must be based on an evaluation of the social capability of the mentally retarded persons by qualified experts, and must be subject to periodic reviews and to the right of appeal to higher authorities.

Above all—the mentally retarded person has the right to respect.

Exhibit 2.1. Declaration of General and Special Rights of the Mentally Retarded (1969).

Normalization

The concept of *normalization* comes from the above rights (Leland and Smith 1974) and has become the guiding principle for programs and services for persons with developmental disabilities. Wolfensberger, one of the originators of this approach, describes normalization as providing mentally retarded and other developmentally disabled citizens with services that are "as culturally normative as possible in order to establish and/or maintain personal behaviors and characteristics which are as culturally normative as possible" (Wolfensberger et al. 1972, p. 28). This means that if we expect people to behave in a more normal fashion and if we provide them with the opportunity and training to do so, they will display more normal behavior and will continue to do so as long as external conditions permit. In short, the normalization principle means making available to developmentally disabled persons daily experiences and activities that are as close as possible to those of the rest of society (Nirje 1969).

How to Apply the Normalization Principle   The normalization principle can be applied to all people with developmental disabilities, regardless of the severity of those disabilities. It can be utilized with all age groups and in all settings. According to Nirje, there are several implications of the normalization principle:

"Normalization means a normal rhythm of day for the retarded." Mentally retarded and other developmentally disabled persons should follow the same type of daily schedule as their normal peers of the same age. This includes getting out of bed and getting dressed daily—even for clients who are profoundly retarded or physically disabled. It means waking up and retiring at normal times, as well as being able to sleep late occasionally. Normalization also means eating under normal conditions—family style, with a small group of people, rather than alone in a room or in a large dining hall. Finally it means having an opportunity for some private time and occasionally being able to break away from the routine of the group.

"The normalization principle also implies a normal routine of life." Think about your daily schedule for a moment. Chances are that you are fairly mobile and move throughout the day from home to work to school to recreation. As you can see, a normal routine includes going to different places during the day to receive services, to earn a living, and to have some fun. Thus, if all of one's life activities occurred in one building or setting, they would not follow a normal pattern of life. This often happens, however, to developmentally disabled persons in institutions or long-term care facilities. All services are usually provided in one setting, and people generally have little opportunity for a change of scenery, let alone a normal life routine.

As part of a normal routine, developmentally disabled persons should be involved in the same types of activities as their peers. This means that school-aged children should attend school and adults, until they reach retirement age,

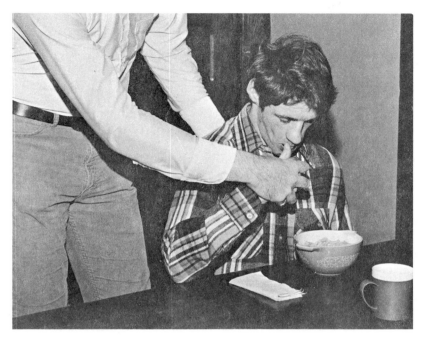

Figure 2.2. Learning eating skills allows a person to be more independent and follow a more normal routine of life.

should be involved in training or work routines. Adults who would not benefit from training or work settings should be involved in daily recreation and leisure-time activities. Outside resources should be utilized to the fullest extent possible for these services.

"Normalization means to experience the normal rhythm of the year, with holidays and family days of personal significance." Holidays, birthday, and special events should be celebrated in a manner appropriate for that person's age. Vacations and trips should be encouraged as a means of refreshing both mind and body. These are all events that add variety and excitement to our lives. Developmentally disabled persons should not be prevented from doing these things simply because of their disabilities.

"Normalization also means an opportunity to undergo normal developmental experiences of the life cycle." Developmentally disabled individuals should have the same personal-growth-enhancing experiences as their normal peers. The more experiences that we have in our lives, the more we will understand our surroundings, ourselves, and others, and the more we will accept our personal strengths and weaknesses. If this is true for normal people, why not also for the mentally retarded and other developmentally disabled people as well?

These people also have the capacity to develop and grow. If they are segregated or prevented from participating fully in life, however, their development will be limited, and they will not achieve all that they are capable of achieving.

With age, the expectations and activities of the developmentally disabled should change. Childhood should be a time of warmth, protection, and shelter, the time of life when one's self-confidence is forming. As people grow older, they tend to participate in more activities away from their primary caretakers. They attend school, become more independent, and spend more time with friends. As they enter adulthood, they tend to move into settings where they can practice their skills of independence. The developmentally disabled, however, might require some special guidance in becoming more independent. Their growth will be limited unless appropriate normalizing experiences are made available. For example, to learn how to relate to one another, people need opportunities to interact. If staff decide in advance that developmentally disabled persons are not capable of forming relationships and provide few opportunities for interaction, the clients will not learn to form relationships. The staff is doing a tremendous disservice to their residents when this occurs.

"The normalization principle also means that choices, wishes and desires of the mentally retarded themselves have to be taken into consideration as nearly as possible, and respected." We all make choices every day—ranging from what clothes to wear to how to spend our leisure time to major life decisions (marriage, children, job). Developmentally disabled persons often have their decisions made for them and are frequently not permitted to exercise even the slightest choice. Lacking appropriate experiences as a result, many developmentally disabled individuals have difficulty making decisions. When this occurs, clients should receive training in how to make simple decisions. After they master this skill, they can learn to make more complex decisions about their lives. Given the opportunity, training (if necessary), and appropriate information, developmentally disabled people can make many sound decisions for themselves.

"Normalization also means living in a bisexual world." Developmentally disabled individuals should be provided with both men and women staff members. Additionally, leisure and work activities should not be segregated by sex. (This topic will be discussed more fully in chapter 10.)

Benefits of Normalization The normalization principle benefits our clients in a number of ways. Most importantly, it assists individuals in developing a positive self-concept. When treated as a person first, an individual begins to experience some feelings of worth and importance. Making decisions (even the most basic ones) about one's life leads to feeling that one has control over oneself and one's surroundings. Having a normal daily routine and expectations for learning fosters growth and the development of new skills.

Normalization and Habilitation The normalized environment includes the habilitation of residents. This means teaching clients to perform many daily activities as independently as possible. At times this might require using adaptive equipment to make the learning process more efficient and the resident more independent. For example, spoons with enlarged handles could be used to help residents with a poor grasp to hold a spoon and feed themselves, and communication boards could allow residents without verbal abilities to communicate. Adaptive equipment thus allows handicapped individuals to participate more fully in the normal routine.

Normalization and Staff The normalization principle also benefits staff, since normalizing the setting for clients simultaneously normalizes the working conditions for you. The rooms and furniture, for example, are generally more pleasant in a normalized setting. Additionally, your role is perceived in a different way. Workers in more normalized settings are regarded as people who can teach clients new skills and assist in enhancing their self-esteem, rather than as mere providers of custodial duties. Staff are regarded as providing important services to important people and as contributing enormously to the habilitation process. Residents' progress toward goals can be viewed as a staff accomplishment as well. Staff can take pride in their clients' learning new skills and in the fact that their person-to-person interactions helped clients to progress.

Normalization and the Developmental Model The normalization principle includes a *developmental model* that states that all people are capable of growth, development, and learning. It must be recognized, however, that merely placing developmentally disabled persons into a normal routine and normal experiences will not always produce the desired effects (Throne 1975). Because of their various handicaps, developmentally disabled individuals will not develop and learn at the same rate or to the same degree as their peers. To learn a new task, for example, a mentally retarded individual may have to repeat it over and over. New tasks might be taught to a developmentally disabled individual by breaking them down into very small steps and teaching each new skill one step at a time. Sometimes other special ways of teaching are necessary. Thus, while residents should follow as normal a routine as possible, they should also be provided with specialized services and training that allow them to develop new skills and move toward independence.

Deinstitutionalization

The basic idea of normalization is that developmentally disabled individuals' pattern of life should follow that of their normal peers. This implies that habilitation takes place best not in large, impersonal institutions, but rather in smaller settings (group homes, intermediate care facilities, nursing care facilities) in the community. Many workers in the field of mental retardation have come to accept this position. As a result, the number of people living in institutions has been gradually decreasing. This process of having people return to the community from institutions is called *deinstitutionalization.*

Deinstitutionalization works only when community resources for the developmentally disabled are available. Housing, transportation, counseling, and job programs in the community are important for two reasons: (1) they help people who are able to stay in the community avoid entering an institution unnecessarily; and (2) they help people leave institutions when they are able to live more independently.

Why Deinstitutionalization Is Important Deinstitutionalization recognizes that large institutions are generally not the best settings for working with people. Whenever large numbers of people live together, some individuals do not receive all of the services that they need. Individuality is frequently lost, since clothing, hairstyles, bedrooms (wards), and daily routines are shared by everyone. These institutions are generally understaffed in professional, direct care, and house-keeping/maintenance personnel. As a result, many clients sit idle for large parts of the day, as there are generally not enough staff to work with them to the degree necessary for their personal growth.

Learned Helplessness A dull environment prevents the residents from experiencing a normal living pattern and reduces their life to a boring routine (deSilva and Faflak 1976). Long periods of time without planned activities and a lack of materials to do things will contribute to boredom. These periods of inactivity can be quite harmful to the clients (DeVellis 1977) because they may become dependent on staff to meet their needs, initiating very few activities on their own and often failing to respond to events in their environment. In short, the clients under these conditions are experiencing a *learned helplessness.* They are becoming unnecessarily helpless because of their living conditions, not their handicaps. These people who have learned not to respond to their environment have a more difficult time learning the skills necessary to master their surroundings. Learned helplessness can occur in any setting where the clients are not provided with stimulating activities and the opportunity to learn new skills.

Problems of Deinstitutionalization The move from the institution to the community is not an easy one, and obstacles are present along the entire deinstitutionalization journey. The first obstacle is within the institution itself: clients must be prepared to reenter society. After living in the institution for a number of years and visiting the community infrequently, many residents do not know the appropriate types of behavior for different settings in the community. Used to having many of their decisions made for them, people reentering society may have problems planning their activities and making many of their own decisions. Additionally, many clients may not have all the self-help skills (dressing, eating, cooking, bathing, use of money and time) necessary to survive in the community. These basic survival skills must be learned before leaving the institution. Without them residents experience culture shock and often fail in their first attempt to move into the community. For deinstitutionalization to be effective, clients must be adequately prepared for their new life-styles before leaving the institution and be linked up with appropriate agencies in the community.

The obstacles continue when the individual enters society. The community is usually poorly informed about developmentally disabled individuals and may unreasonably fear interacting with them or oppose the establishment of new residential settings (group homes, intermediate care facilities) or training facilities (schools, workshops). Social service agencies, already overloaded with other programs, may find it difficult to add the necessary services. In many locations appropriate services for developmentally disabled individuals are lacking, both in terms of quantity and quality. In larger cities there are usually not enough residences (group homes, foster homes, and the like) or work and leisure-time settings for all the people expected to live in that community. Mental health services for the developmentally disabled are often scarce as well, and when services are available, they often fail to address the specific needs of the clients. Poor quality services often result from the lack of awareness and training of many professionals and paraprofessionals with regard to the needs of the developmentally disabled.

These last obstacles to deinstitutionalization can have a negative effect on the clients, many of whom return to the institution because there are not enough normalizing alternatives available in the community (Conroy 1977). This does not mean, however, that the appropriate services cannot be provided in the community. In fact, more and more people are becoming aware of the need for community services for the developmentally disabled. They are developing and improving more resources to make sure that quality services are being provided. This encourages more people to pursue careers in the fields of mental retardation and developmental disabilities, thus enriching clients' lives and their prospects for quality care in the future.

As the services improve both in quantity and quality, the likelihood that people will be able to remain in the community will increase. Fewer people who were "deinstitutionalized" will return to the institutions. Also, fewer mentally retarded persons who have not previously entered the institution will need to, as they will have the services they require in their own communities.

COMMUNITY SERVICES

Services to the developmentally disabled should meet the individual's needs through his or her entire life span. Resources must be made available to young children living with their parents, as well as to adults and the elderly residing with their families, independently, or in some supervised setting. These cradle-to-grave services in the community should include residential, vocational, educational, recreational, mental health, and other support systems. Let us explore some of these different types of resources which are becoming more and more common in many communities.

Residential Programs

Group Homes   Group homes (hostels, halfway houses) and apartments have become quite popular over the last several years. These generally house from two to ten residents and have been established for children as well as adults with a variety of handicapping conditions. The amount of supervision and guidance varies from home to home, depending on the clients' needs. People in group homes usually work or go to school in the community and can make use of all activities and resources available to the general public. Group homes can serve a variety of functions. They can be used as transition points where people leaving the institution or their families can learn the skills necessary to live independently or semi-independently. Group homes can also serve as long-term residences for people who are comfortable there or who do not have the skills necessary to live without some form of supervision (Robinson and Robinson 1976).

Nursing Care Facilities   Nursing homes, primarily intermediate care facilities, are additional residential centers for developmentally disabled people who are aged or require continued medical treatment. Until recently there has been very little emphasis on the development of specialized centers for the aged retarded. As a result, these persons have entered nursing homes with the elderly from the normal population. This often works out well if the staff and administration recognize the developmentally disabled's special needs and assist them in learning new skills.

Specialized nursing homes called *intermediate care facilities for the mentally retarded* (ICF/MRs) provide health care and habilitative services to individuals who need physical, emotional, social, and educational support not available in their homes. Residents of ICF/MRs may have severe physical, adaptive, behavioral, and learning problems.

Foster Homes   Foster homes and family care are other residential alternatives to the institution. These settings are especially important to the developmentally disabled of all ages, who can be provided with much individual attention in a home environment. Foster homes have become much more feasible as the other community resources (schools, workshops, recreation) have become more

Figure 2.3. Group homes in the community are a popular alternative to institutional living.

available and able to provide the resident and the family with the necessary support to allow the individual to reside in the community.

Respite Care Improved community services have also allowed many parents to raise their developmentally disabled children at home, an arrangement that of course follows the normalization principle. Occasionally, however, parents need a break from the daily routine and require some time alone. *Respite care*

Figure 2.4. Living with others in a community setting means that everyone is going to have to pitch in and help.

homes have been developed to respond to such needs. The difficulty of finding babysitters for the developmentally disabled children often interferes with parents' social and personal life and creates a great deal of strain within the family. With respite care homes, parents can make arrangements to leave their child for a weekend or a few days, knowing that he or she will be well cared for. In this way family balance can be maintained, and there may be less desire to place the child outside the home on a long-term basis.

EDUCATIONAL SERVICES

Educational services for developmentally disabled children should start early as a way of preventing or slowing down further delays in development. These *early intervention programs* take a variety of forms, including infant stimulation programs and early training classes. In infant stimulation programs a teacher visits the home and teaches the parents methods of stimulating and working with their child. In early training classes, the young child attends a developmental classroom setting with several other children, often while the parents are involved in some educational program or support group. Preschools are available for many three- and four-year-olds that stress language development, gross motor activity, and socialization.

Recent federal legislation has mandated the right to education for all children, regardless of handicap. As a result, no child can be excluded from the educational process, and schools are developing new programs to meet the needs of all developmentally disabled school-aged children. As part of the school program, prevocational training should be included for high-school-aged students to provide them with a work orientation and help them understand that they can be productive members of society.

VOCATIONAL TRAINING AND SHELTERED WORKSHOPS

When these students complete their school programs, they should have the opportunity for further vocational training, and with appropriate instruction many will be able to work competitively. Others will require more supervised, less competitive employment. To meet this need, *sheltered workshops* have been established, where mentally retarded and physically handicapped individuals can work productively. These workshops have had a very positive effect on many developmentally disabled persons, who come to view themselves as important. These individuals produce worthwhile products, have the opportunity to learn new skills and to be with people, and earn a paycheck.

Other developmentally disabled adults may require more training in appropriate social as well as work behaviors. Work activity centers have been set up to meet these needs. As these people learn new skills, they often move on to other job training sites or to jobs of their own.

OTHER SERVICES

Since many mentally retarded and other developmentally disabled persons have difficulty organizing their free time, it is very important that they have a chance to participate in structured recreation and leisure-time programs. At times some individuals will have to be taught how to plan or choose between activities or how to participate in them (bowling, dancing, cooking). Some people might also need to learn how to interact with others during the recreation program.

These recreation and socialization programs should occur during nonwork hours (evenings and weekends) for those people who are employed. But what about those who are not employed, such as the severely handicapped, the elderly, and the emotionally disturbed? None of these groups would benefit by merely sitting idle day after day, but all would benefit from out-of-home experiences (Robinson and Robinson 1976). For these people daytime activity centers where they would have the opportunity to be with others, to learn new skills, and to have some purpose to their day would be helpful.

At different times throughout the lives of developmentally disabled persons and their families, there are crisis points where counseling and mental health services could prove helpful. Many mentally retarded adults, for example, have difficulty understanding new roles and expectations that others have of them as they grow older. Some have difficulty accepting their disabilities or breaking away from their families. Still others might have deep-seated emotional problems that require attention. Parents too have difficulties that result from having a child with a disability and often need guidance in childrearing techniques, suggestions for placement in a group home, or help in working through personal problems that the developmentally disabled child's presence has intensified.

QUESTIONS

2-1. List the number of ways that you think that the developmentally disabled are like everyone else.

2-2. Do you think that if we followed up on Dr. Goddard's plan and totally isolated mentally retarded persons from the rest of society, that we would eventually have no mentally retarded people? Why or why not?

2-3. Are there dehumanizing experiences in your work setting for your clients? Can you think of ways to improve the situation?

2-4. Obtain a copy of Blatt and Kaplan's *Christmas in Purgatory* and pass it around the class. What are some of your reactions?

2-5. Discuss the relevance of the normalization principle to your work setting. How can you make life more normal for your clients?

2-6. If you were trying to start a group home for mentally retarded adults, what arguments would you use to convince neighbors to allow the group home to be established?

ACTIVITY FOR THOUGHT AND DISCUSSION

View the film *Normalization: The Right of Respect* (available from Atlanta Association for Retarded Citizens, 1687 Tully Circle N.E., Suite 110, Atlanta, Ga. 30329).

Developing Habilitation Programs for Clients: The Interdisciplinary Team 3

As discussed in the previous chapter, one of the current thrusts in the field of developmental disabilities is habilitation. The habilitation process is very important; it must be well organized and preplanned, involve each client in programming, and chart each client's progress. How do we develop programs, how do we write programs, and how do we keep track of them all without getting totally confused? The idea of the Individual Habilitation Plan (IHP) has been developed in response to such questions.

THE INDIVIDUAL HABILITATION PLAN

The IHP is composed of

1. A statement of goals and objectives for each client, developed jointly by all the people working with that client.
2. A plan of techniques and methods to achieve the objectives.
3. A list of people who will be responsible for carrying out the program.

Usually, the team members (all the people working with the client) develop the IHP after a period of observation and evaluation of the client. The team then uses each member's observations in determining the client's strengths and needs, so that they can both be included in the IHP. Once the plan is written and carried out, the client's progress is measured through data-keeping procedures. The plan and the client's progress are also reviewed by the team on a regular basis (every thirty or ninety days), and the plans are revised as necessary.

The IHPs are important not only in terms of organizing everyone's efforts in working with the client, but also in terms of providing feedback, both to the client and to the staff. Learning that people are working with him or her often encourages the client to be more active in the program. If successful, the program provides the client with new skills that allow more independence. It also provides an ongoing record of the client's progress and accomplishments. Similarly, IHPs are helpful to staff, providing them with specific duties in working with each of their clients and allowing them to work constructively even with difficult clients. Also, the IHPs are a means of keeping track of what the staff had accomplished with each client (Houts and Scott 1975, Parham, Rude, and Bernanke 1977).

THE INTERDISCIPLINARY TEAM

As stated above, the IHP is developed jointly by a team of people, each of whom has been working with the client. This is called an *interdisciplinary team,* in that all the disciplines (e.g., medicine, nursing, direct care, psychology) are planning the client's programs together to develop some common goals and directions. Each member of the team brings in his or her expertise and knowledge of the client and shares it with the other team members. Then all this information is put together to develop one plan for the client.

While each client has his or her own team of workers, because of their specific needs some clients might have larger teams than others. Thus, one client's team may be made up of direct care workers, a nurse, and a social worker, while another client might have in addition a psychologist, a physician, and a recreation worker. One discipline that will be represented on every team, however, will be the direct care staff. (Some states and some facilities require a core group of disciplines on each client's team. Check with your facility to see if such a requirement exists.) At times, additional information about the client may be helpful in developing the IHP, or a different point of view might be welcomed. In such cases other staff members may be asked to join the team as regular members or as consultants. Let us examine what the different disciplines have to offer to the IHP process.

Direct Care Staff

In various facilities direct care staff are also known as hospital aides, nurses' aides, developmentalists, and trainers. (Throughout this book, we will use the term *direct care staff.*) These workers are usually most familiar with the client and can usually define the client's likes, wants, and needs. Direct care staff can also describe the client's interactions with other residents and staff and progress in existing programs. Finally, they can report on the client's self-help

Figure 3.1. This is an interdisciplinary team meeting in progress.

skills, communication skills, recreation skills, and other important behaviors and abilities. Direct care staff are crucial team members: they interact most with the client, have important information to contribute to the habilitation plan, and will be responsible for carrying out many of the programs.

Physician

The physician is frequently involved in the habilitation planning process as well. The role of the physician is not to direct the team nor to provide guidelines within which the team is to work. He or she is an expert, however, in determining the overall physical health of the resident and should report this information to the team. The physician should also review with the team any medical treatments that involve the client, so that the entire team is familiar with the client's medications and their side effects. Additionally, the physician is responsible for making sure the resident receives appropriate diet, health care, and psychiatric care.

Nurse

The nurse is often a member of the habilitation team and may present the physician's report and recommendations in the latter's absence. The nursing staff member also has important responsibilities of his or her own for the day-to-day health and safety of the resident. The nurse may make recommendations about the physical care and habilitation of the resident, as well as the ongoing medical care. The nurse may also be responsible for supervising programs carried out by the direct care staff and may suggest ways that the programs can be written to meet the client's physical limitations.

Physical Therapist (PT) and Occupational Therapist (OT)

The PT and the OT are concerned with enabling the client to perform tasks that increase his or her control over the body and the environment. This is especially important when the client has a physical handicap or a brain dysfunction. Recommendations from these professionals may include exercises and activities that encourage the client to use muscle groups that are not functioning properly. The PT and the OT can also assist the habilitation process by suggesting activities to improve the client's social and self-help skills. These recommendations might include the use of adaptive equipment (e.g., wheelchairs, special spoons and plates) that would allow the resident to be more self-reliant.

Instructor

If the client is involved in any educational or vocational program, the teacher or vocational instructor should be a member of the team. These people can provide information about the client's abilities in the following areas: gross and fine motor skills, academic or preacademic skills, self-help skills, language and communication, and socialization. These professionals can also provide information about vocational or prevocational skills, budgeting, concepts of numbers, time, and money, and other daily living skills.

Psychologist

A psychologist may be asked to evaluate the client and provide information to the team, such as the client's level of intellectual functioning or ability to learn. He or she may also recommend the best techniques to use in teaching the client a new skill and evaluate the client's adaptive behavior (how well the client is coping with the demands of the environment). The evaluation of adaptive behavior may include self-help and communication skills, emotional and social behaviors, self-direction, and effective use of time. The psychologist can then make recommendations for programming in each of these areas, as well as provide direct services to the client in terms of training in new skills and counseling.

Social Worker

A social worker may also be a member of the team when contact with other agencies, parents, or friends is necessary. The social worker is responsible for making certain that the client is receiving appropriate services and benefits, both within the facility and from outside sources. This team member also looks after the client's personal interactions with parents, relatives, and friends. At times, the social worker may act as the client's advocate, making sure that the client's needs and desires are brought into the team meeting. He or she may also provide services to the client and the family, including counseling, contact with other agencies, and arranging for appropriate services.

Recreational Therapist

A recreational (activity) therapist may be invited to attend the team meeting, or to submit a report, if the client is participating in recreational programs. The recreational therapist can teach the client to use leisure time effectively by participating in games, interacting with other people, and playing and having fun. The recreational therapist can also help the client gain better control over his or her body through specialized exercises, swimming, and other experiences adapted to meet the client's needs.

Client

Last and most important, the client and his or her family should be active members of the team. Whenever possible, the client should attend the team meetings and have direct input into the development of the habilitation plan. When this is not possible, the goals of the habilitation plan should be explained in advance to the client. This is important and should *always* be done, whether or not it is felt that the client can understand the goals.

In reviewing the roles of the different disciplines, you will note that there is much overlap of responsibility and knowledge. This common ground allows the team to plan joint goals for the client and to develop a common direction. The IHP also promotes open communication between the team members.

Developing the IHP 4

THE STRENGTHS/NEEDS LIST

Once all the team members have delivered their reports, we use all this information to develop a list of the client's strengths (abilities) and needs (areas to be improved). This list provides the basis for developing the IHP and its goals.

It is usually easier to identify a person's weaknesses and needs than his or her abilities and strengths. It is important, however, to force ourselves to think of the client's strengths first, as this is more difficult, and then think of the areas that the client should be working on (needs). Once this strengths/needs list is developed, we try to use as many of the client's strengths as possible in helping meet needs (Houts and Scott 1975).

Strengths include

a) All those things the client does well (tells time, rides the bus, does laundry)
b) Things the client enjoys doing (bowling, listening to music)
c) Things the client enjoys receiving (rewards, soda pop, games)
d) Special events the client enjoys (parades, parties, birthdays)
e) Positive personality characteristics (prompt, reliable, eager)
f) People the client enjoys being with (staff, parents, friends)
g) People that enjoy being with the client

Needs include

a) Skills the client wants to acquire (to tie shoes, to bathe independently)
b) Skills the client wants to improve (to eat more neatly)

    c)   The elimination of social/vocational/academic problems that are interfering with personal adjustment (to get along with others)

    d)   Other skills that the staff feels the client should learn

Let us look at an example and see if we can develop a strengths/needs list:

John is a twenty-four-year-old, severely retarded man who has some problems in controlling his arms and legs. He participates in all activities that staff ask him to join, but does not initiate many activities on his own. John's parents visit him frequently and take him out for walks and drives to nearby places. John finger-feeds, but does not use table utensils. Mr. Caring, his staff member, makes sure to spend time with John every day, and John looks forward to these visits. John is friendly with all residents and staff; he enjoys special events and swimming. John is able to get around the building on his own, but often stops to talk to staff in their offices.

A strengths/needs list for John would be as follows:

Strengths	Needs
1. Participates in activities when asked.	1. Needs to learn to initiate his own leisure-time activities.
2. Visited often by his parents.	2. Needs to learn to use table utensils (spoon, fork).
3. Enjoys spending time with Mr. Caring.	3. Needs to recognize when he is interrupting others.
4. Cared about by Mr. Caring.	4. Needs to know when he can stop and talk, and when he has to go to his assigned place (activities center, dining room)
5. Can get around the building on his own.	
6. Is friendly.	
7. Can finger-feed.	5. Needs OT/PT consult for ideas about improving gross and fine motor skills.
8. Likes special events.	
9. Likes swimming.	
10. Likes to talk to staff.	

ACTIVITY FOR THOUGHT AND DISCUSSION

Now that you know what strengths and needs are, break up into pairs and do the following in class:

1. First person in pair: Think of a client, and with the assistance of your partner, develop a strengths/needs list. The second person should ask questions to make sure that all of the points covered above are included in the list.

2. Arrange the list of needs in order of importance.

3. When you are finished with the first person's client, reverse the roles and do the same thing for the second person's client.

4. When you have finished, review your lists with the whole class and discuss how you would use the clients' abilities to help meet needs.

(Note to the instructor: During this step, make sure that the students' use of clients' strengths is realistic, practical, and conforming to the normalization principle as much as possible.)

GOALS

After assessing and listing all the client's abilities and areas of need, the team decides on some general directions in which the client should be moving. These statements of the general direction of the program are called *goals.* Examples of goals for clients might include the following:

John will be more responsible for his personal hygiene.

John will interact cooperatively with others.

These are rather vague statements, but they do describe what the client will be expected to do as a result of the training. Because they are so general, goal statements do not provide enough specific information to help staff in programming for clients. Therefore, IHPs must also include objectives and methods for achieving the objectives, and ultimately, the goals. (Some authors use different labels for goals and objectives, such as *long* and *short-range goals,* while others use them interchangeably.) In addition, the IHP should state a target date for achieving each objective and specify who is responsible for each objective. A typical IHP form is shown in Exhibit 4.1.

OBJECTIVES

The objectives section of the IHP form describes what the client will be doing after completing the training program. The *methods* section describes what the

Individual Habilitation Plan

Resident's Name _____ Date _____

Case Manager _____

Team Members Present _____

_____

Goal Statement _____

Objectives	Methods	Date Started	Target Date	Completion Date	Provider

Exhibit 4.1. Sample IHP form.

staff will be doing to help the client to meet these objectives. Specific techniques for the staff will be discussed later in the context of *behavior modification* (chapters 5 and 6). For now, let us talk about objectives. These are important to the client's progress, and must be *stated* clearly, since they are supposed to describe exactly what the client *will be doing* at the end of training.

Writing in Behavioral Terms

Because the objectives must describe the client's behavior, they should be stated in behavioral terms. Behavioral statements are characteristically *observable, measurable,* and *understandable.*

Observable The statement must describe something that we can see happening. Thus, we do not describe the person's emotions or thoughts, but rather his or her behavior. We can talk about the client's hugging, kissing, and smiling, since we can see these behaviors (Golden and Ho 1973). The objectives should also include how and when the behavior is to occur: the conditions (after dinner, while eating), the situation (at work, at home, in the movies) and the people involved (with the staff, with the client). These help make the objectives more observable.

Measurable We must be able to count or measure the behavior in some way. For example, we cannot count *likes* or *loves,* but we can count *smiles* or *hugs.* The objective should also include the number of times that the client must perform the desired behavior before it can be considered successful (five times, five out of ten times, 80 percent, five out of seven days, 100 percent).

Understandable Anyone should be able to pick up your IHP, read it, and understand it. It has to be written that clearly.

QUESTIONS

To the instructor: You may use the following as a class exercise by writing the statements on the blackboard and asking the students to respond. Provide feedback to the class using the answers provided in appendix A.

4-1. Are the following objectives written in good behavioral form? Improve them using the three criteria for a behavioral statement just discussed. Remember, describe what the client will be doing to achieve the objective.
   a) John will dress himself.
   b) John will interact with others.
   c) John will learn to tell time.
   d) John will be taken to the dining room daily.

(Turn to appendix A for examples of these objectives rewritten in behavioral terms.)

4-2. For additional practice in writing clear behavioral statements, change the following ideas into clear behavioral statements:

a) Does more on his own.

b) Friendlier to others.

c) Talks more.

d) Acts appropriately.

Ask your instructor to review your behavioral statements and provide you with feedback.

TASK ANALYSIS

The idea behind the habilitation plan is to have the client achieve the goal. Often, however, the goal is too large to tackle all at once, and trying to do so usually leads to failure for the client and the staff as well. To overcome this problem, we must develop a step-by-step procedure to reach the goal. This is called a *task analysis*—breaking down the task into easily obtainable steps.

Figure 4.1. Many household chores can be taught by a task analysis so that residents may learn to do things efficiently.

Figure 4.1 (*concluded*)

Helping the Client to Succeed

The goal has to be broken down into small steps so the client can succeed. In this way the client can work on a step until mastering it and then move on to the next step. Working one step at a time, the client will eventually master the entire goal. By breaking the program into smaller steps, we make it possible for the client to succeed and to gain self-confidence. We also provide a means for staff to see the client's progress from step to step, often increasing staff motivation. As staff see the client progress, they begin to realize their importance in the client's program.

When breaking a goal down into workable steps, *each step must take the form of a behavioral statement.* Thus, when writing a task analysis, keep the three criteria for behavioral statements in mind. Behavioral statements must be observable, measurable, and understood by all who read them.

Once the goal is selected and the task analysis written, the program is carried out. If you look at the data and decide that a program is not working, one of the things to examine is the task analysis. Look at each step and ask yourself: Is it too difficult for the client? Does it have to be broken down further?

An example of a task analysis is the following program to teach hand-washing skills (we will discuss the specific techniques used to teach each of these steps later). If such a program is used, and the client has difficulty with any of the steps, break that step down further. For example, in the following program, step 3 (client will take hands out of the water and pick up the soap) can be broken down into two steps: Client will take hands out of water. Client will pick up soap.

Goal: *The client will wash hands when told 90 percent of the time.*
1. Client will turn the water on.
2. Client will put both hands in the water.
3. Client will take hands out of the water and pick up the soap.
4. Client will rub the soap between hands until they are lathered.
5. Client will return soap to soap dish.
6. Client will rinse soap from hands.
7. Client will dry hands.

In teaching the client a new skill, the steps of the task analysis are taught one at a time. After learning to do one step, the client should be expected to keep on doing it while learning the next step in the sequence. At first, the instructor might have to guide the client physically through the step (take the client's hands and guide them) and demonstrate or tell the client what to do. When the client gets to the point of doing a step with little or no direction, it is time to move on. Again, if the client has difficulty with any of the steps, break them down into smaller segments.

ACTIVITIES FOR THOUGHT AND DISCUSSION

A. By yourself or in a small group, select one of the following goals and do a task analysis. First, write a clear behavioral statement of what you expect the client to be doing (objective). Then, break that task down into four or five smaller steps (or as many steps as needed), each written in behavioral terms. After completing the program, turn to pages 91, 113, and 68, for examples of task analyses for the following programs:

   a) Showering
   b) Putting a four-piece puzzle together
   c) Putting on pants

B. View the Marc Gold film *Try Another Way* (may be rented from Try Another Way, 128 E. 36th Street, Indianapolis, Ind. 46205).

Behavior Modification: Teaching New Behaviors 5

The purpose of behavior modification is to teach new skills to people so that they may become more independent, and to reduce or eliminate behaviors that are harmful to themselves and others. These changes in the client's behavior are not accomplished haphazardly but rather in an organized and preplanned manner. It should also be stressed that programs to eliminate inappropriate or harmful behaviors should always teach more appropriate or useful skills. For example, when attempting to reduce the number of times that clients slap themselves, the instructor should also teach the clients to do something more appropriate with their hands (e.g., clapping, throwing a ball).

OBSERVATIONAL SKILLS

Before an IHP is developed with a client, there must be some information upon which to base the programming. One way of assessing or gathering information about a client is through using good observational techniques. In carrying out behavior modification programs, we will be concerned about behavior, and therefore we have to observe and measure it accurately. There are several different ways of doing this, and we will want to select methods that are quick and easy to use, yet provide all the information we need.

The Anecdotal Log

If there is some question as to what behavior to observe, and more information is needed about the client, a simple *log* might provide this information. The log should include both positive and negative behaviors and is usually written out in anecdotal form, as the behavior occurs or at the end of the shift. The log can help in communicating with staff on other shifts and in getting a general idea of what the client is doing. It is not, however, always the best method of observation and recording. Details can be forgotten before the staff has time to record them and other important information might also be omitted, such as events just before or just after the client's behavior. For example, a log completed at the end of a shift might include the following report on one person:

Name: John Client

John acted up all day. I do not know what is bothering him, but he hit and pushed other people constantly. John spent a few minutes this morning working on a project, but then threw it against the wall for no reason. His instructor helped him put it together again.

This log gives us some information, but not a time-frame for the events or an accurate order of all the events. And it fails to report what happened before or after John acted in each of these ways. To do programming to help John, staff must have more detailed information. (See Exhibit 5.2 for a log that provides more information than the usual anecdotal log.)

Modified Log

There is a more workable recording method than a simple log. With this modified log (see Exhibit 5.1) we begin by dividing the paper into three columns, marked *Time, Resident's Behavior,* and *Before and After.* All the subject's behaviors are recorded in order in the second column. Any actions directed toward the subject by other people, either before or after the resident acted, are recorded in the Before and After column and appear right across from the subject's behavior. The time of each event is marked in the first column.

The modified log in Exhibit 5.2 provides us with much more detail than the anecdotal log about the client we have discussed. It tells us more about the order of events and suggests possible cause-and-effect relationships. This additional information assists staff in programming with the client. For example, from this modified log it can be seen that as long as John is getting staff attention, he is fairly well behaved. On the other hand, John will also act up (hitting, throwing, turning the TV on loud) to get staff attention. Also, it is clear that John did not act up all day, but in fact had some good moments. This additional information gives the staff clues as to how to respond to John.

Frequency Count

If you know *exactly* what behavior you want to observe, then other observational methods might be quicker than the modified log. One alternative is a

MODIFIED LOG

hr/ Time	Resident's Behaviors	Before & After
	Date	Observer	Resident _____ hr

Exhibit 5.1. Modified log form.

Time	7-5-78 Date	Robert Goodsight Observer	John Client Resident
	Resident's Behaviors		Before & After
7:05 A.M.	John lying in bed. John hit roommate as he was passing by.		Roommate got up to go to bathroom. Roommate yelled and staff responded by talking to John about hitting others.
8:15	John on way to breakfast pushed another client.		Client pushed back; staff separated John and talked to him about pushing.
9:00	John attended recreation program. Worked hard on project.		Instructor spent time with John and assisted him with the project.
9:10			Instructor working with another client.
9:12	John threw his project against the wall.		Instructor ran over to John and helped him put project together again and calmed him down.
11:30	John and another client playing cards—no problem.		
12:00	Lunch—no problem.		Other clients talking to John.
1:00 P.M.	No scheduled activity. John alone —no problem.		
1:10			Roommate returns from lunch.
1:15	John turns his TV on loud.		Roommate complains to John and then goes to staff. John told to turn TV down.
2:00	John in room; is doing well—talking to nurse and having a cup of coffee.		Nurse spends a few minutes talking with John.
2:10			Roommate returns to room and starts talking to nurse.
2:10	John hits roommate and tells him to be quiet.		Nurse calms John down and leaves.
2:45	John says goodbye to staff who are leaving and hello to new staff.		Staff talks to John.

Exhibit 5.2. Sample modified log.

frequency count, that is, counting the number of times that a specific behavior occurs within a period of time. Examples of frequency counts are

the number of times a client gets out of his chair during dinner

the number of questions answered correctly by a client within ten minutes

These both involve counting specific behaviors occurring within a specific amount of time (during dinner or ten minutes).

How do you record behavior using a frequency count? Simply count how often the behavior occurs. You can use a tablet and pencil and put a slash mark down each time a behavior happens. Include the times when you start and stop your observations. An example is provided in Exhibit 5.3:

John's Out-of-Seat Behavior

Start   9:45
Stop 10:05   H̶H̶t̶   II

Exhibit 5.3.   Simple frequency count.

In twenty minutes, John was out of his seat seven times. What behaviors lend themselves to frequency counts? Obviously they have to be something that you can count. They have to have a very definite beginning and ending and should not run into or be confused with other behaviors.

If you are going to be observing an individual for a long period, you might find it easier to record behaviors in shorter blocks of time. This will provide additional information concerning any changes of behavior with time. For example, if you are observing the number of tasks completed by a client for an entire day, you might want to break your observations into fifteen-minute blocks, as in Exhibit 5.4.

Time	Number of Events
9:00	I
9:15	H̶H̶l̶
9:30	III
9:45	H̶H̶l̶ I
10:00	H̶H̶l̶ III
10:15	II

Exhibit 5.4.   Frequency count using short blocks of time.

You now have the additional information that the client completes more tasks between 9:45 and 10:15, and you might explore possible reasons for this increased productivity, which might be applied to the other time periods.

Time Sampling

The frequency count is very quick and easy to do, but it does not record before-and-after behaviors, so we cannot know the consequences of the client's behaviors. Also, this method is less effective for observing more than one behavior or more than one person. When we need more information than the mere number of times that a behavior occurs, we must use a different observational technique. One such method is *time sampling*. With this type of observation we break the day (or any length of time we are observing) down into equal parts (e.g., five minutes, ten minutes, half an hour). If more than one person is going to be observed (e.g., John and others from the modified log in Exhibit 5.2), we simply divide the horizontal rows by the number of people to be observed. If there is more than one behavior to be observed, then we use a coding system.

Using the example of John in the modified log, we can fill in the time sampling form in Exhibit 5.5. This technique provides a lot of good information quickly. It allows us to keep track of more than one behavior at a time and permits us to observe more than one person. Thus, this technique allows us to understand the effect of one person's behavior on another. Time sampling is a good observational technique, but it should only be used when it is known in advance what behaviors will be observed.

QUESTIONS

5-1. Do a modified log on a client for one day. Bring it in to share with the class. What are some of the advantages and disadvantages of this technique?

5-2. Pick one or more behaviors from the modified log report you completed. Use either a frequency count or time sampling procedure to collect additional information on your client. How can you use this additional information in programming for your client?

TEACHING NEW BEHAVIORS

Reinforcement

One of the tools often used to teach new behaviors is the task analysis that we have discussed as part of the IHP written under the heading *Objectives* (see pages 43–45). One procedure that staff can use to help clients achieve those objectives is to reward them whenever they do something the right way.

	10 min.	10 min.	10 min.	10 min.	10 min.	10 min.
7:00 John	H					
Other	Y (T)					
8:00 John		P				
Other		P (T)				
9:00 John	G	Th				
Other	(T)	I (T)				
10:00 John						
Other						
11:00 John			G T			
Other			T			
12:00 John	G T					
Other	T					
1:00 John	G	A				
Other		T (T)				
2:00 John	G	H			T	
Other	(T)	I (T)(I)			(T)	

Staff's behavior

(T) = staff talking   (I) = staff ignoring

Patient's behavior

H = Hitting   P = Pushing   A = Annoying behaviors (e.g., turning TV on loud)

G = Good behavior
T = Talking   Y = Yelling
                                           I = Ignore

Exhibit 5.5. Time sampling coding system.

Rewards are provided because it is believed that people learn to behave in certain ways because of the consequences of their behavior. If something they perceive as "good" happens after they behave in a certain way, people will tend to repeat that behavior. Another word for reward is *positive reinforcer,* which can be defined as anything that is presented immediately after a specific behavior that makes the behavior more likely to occur in the future.

Unlearned Reinforcers

There are many different types of reinforcers, and different reinforcers have different effects on different people. Some things are natural reinforcers, that is, things that a person does not have to learn to like. Some examples of *unlearned reinforcers* are food, water, sex, and warmth, which are usually very powerful reinforcers and can maintain behaviors (keep behaviors going) for a long time. Out of this group of reinforcers, food is most commonly used to teach new behaviors, especially with people with severe and profound mental retardation.

Learned Reinforcers

Another group of reinforcers are things that people learn to enjoy. A good example of a *learned reinforcer* is money. If you give a baby a nickel, it means nothing to him or her, except as something to roll or throw. But as it becomes paired with other things (e.g., buying candy), the nickel soon begins to have value of its own. In order to keep its value, a learned reinforcer must occasionally be paired with other reinforcers. Thus, as long as we can buy things with it, money will remain a powerful reinforcer. But if the opportunity to exchange money for desired objects becomes rare or nonexistent, then money becomes valueless and loses its strength. For example, if you are deserted on an island with a million dollars, the money will slowly lose its importance, as long as it cannot be exchanged for desired objects.

Some other examples of learned reinforcers include social rewards like praise, smiles, and pats on the back. The goal of behavior modification is to get the behavior under the client's control, maintained by natural reinforcers (reinforcers that occur, unplanned, in the environment). Therefore, we eventually want to get away from the unlearned reinforcers like food and have the behavior maintained by social reinforcers like praise. Some clients, however, do not respond to praise. For these clients food (or other reinforcers) and praise might be presented together until a strong association is formed between the two. Then, praise would be continued as a reinforcer while food would be phased out. In this way the praise becomes rewarding on its own simply because of its association with the food.

Other things can be reinforcers as well. Activities, music, games, special events, quiet times—almost anything can be used as a reinforcer, as long as the client is willing to work for it. A special event (going to the movies or some other recreational activity) can be set up to occur only after the client performs appropriately, according to his or her habilitation plan. Regardless of the type of reinforcer used, however, in setting up programs with clients, we must start them at or below the level at which they are functioning. This is important in motivating the client to perform, in that it provides success experiences, which encourage him or her to continue the behavior that brought success. The client also learns to do what is expected and progresses as staff increase the requirements that the client must fulfill before being reinforced.

USING REINFORCEMENT EFFECTIVELY

Reinforce Immediately

Reinforcing the person immediately (within half a second) after completing the desired response is generally more effective than after some time lag (a few minutes, hours, days). You can see by this why praise is usually such an effective reinforcer. Praise is always available and can be given quickly when the client makes an appropriate response.

Amount of Reinforcement

The amount of reinforcement to give the client for correct responses is also an important issue. If food is used as a reinforcer, very small amounts are usually more effective than large amounts, especially if the client will have an opportunity to perform correctly many times in a short period of time. With small amounts the client will become tired of the food much more slowly than with larger amounts and will remain motivated to participate in the activity for longer periods of time. Also, if a particular food is used, the client should have gone without it for some time before the training session. If ice cream is being used as a reinforcer, for example, it would be best if the client did not have ice cream immediately before the training session in the interest of client motivation during the session. If such a situation occurs, switch to another reinforcer. Also remember that whenever food is used as a reinforcer, it should be accompanied by praise.

If social reinforcers are being used, it is important that they not be too weak or too strong. Being gushy, or overly excited or insincere about the client's performance, can grate on everyone's nerves if done often enough. On the other hand, not providing enough enthusiasm might fail to communicate the desired message to the client ("You did well," "I am proud of you," etc.) and therefore, might not be effective as a reinforcer.

It is also important to provide novel reinforcers occasionally. This way, the clients will not get tired of "the same old thing" and will tend to remain motivated for longer periods. Also, providing the client with choices of reinforcers makes them more powerful because the client will select what he or she wants most.

QUESTIONS

5-3. List all of the types of reinforcers that you think would be helpful to your clients.

5-4. Do you think that a client needs only one effective reinforcer, or should a variety of reinforcers be made available? Why?

5-5. Do you think activities can be reinforcers? List some that can be used in your work setting.

OTHER TRAINING TECHNIQUES

Several other techniques can be used in conjunction with task analysis and reinforcement. Several such techniques are used when the client is unable to perform even a part of the desired behavior. For example, a profoundly retarded client who cannot eat with a spoon can be taught this self-help skill without too much difficulty, using a program similar to the following:

SPOON-FEEDING

Goal: *John will use a spoon to feed himself.*

Objectives	Procedures
1. John will eat with a spoon when guided 95 percent of the time.	1. a) Stand behind John and place a spoon in his hand. b) Guide John's hand and spoon into food. c) Lift John's hand and spoon to his mouth. d) Praise John and repeat process.
2. John will be able to guide a spoonful of food to his mouth 95 percent of the time.	2. a) Stand behind John and place a spoon in his hand. b) Guide John's hand and spoon into food. c) Lift John's hand and spoon until spoon is about to enter his mouth; then release hand. Praise John and repeat. (When John can get the spoon into his mouth consistently, proceed to d.

60

	d) Lift John's hand and spoon until it is halfway to his mouth; then release hand. Praise John and repeat. (When John can do this consistently, proceed to *e*.)
	e) Help John place food on the spoon, and then let him take the spoon to his mouth.
3. John will be able to scoop his food with a spoon and guide it to his mouth 95 percent of the time.	3. Assist John in scooping food, allowing him to do more and more on his own, until he is eating independently.

If the client has difficulty grasping the spoon or scooping the food, consult your occupational therapist, so that adaptive equipment (a spoon with a bent or enlarged handle or a plate with a high ridge to scoop the food against) may be obtained.

Shaping

In this spoon-feeding program food is the more powerful reinforcer, with praise being used to provide additional feedback to the client. One of the techniques described in this program is called *shaping,* which involves reinforcing the client for doing more and more of the desired behavior until eventually the goal is reached. To use shaping effectively, you should

1. Start with a response that the individual can already give (eating the food from the spoon).
2. Reinforce small changes in that behavior as it gradually becomes the goal behavior (moving the spoon to the mouth and scooping food).
3. Know exactly what behavior you want. The goals *must be* well defined in advance, as well as the individual steps that lead to the goal. A well-defined task reduces the chance of reinforcing undesired responses and increases the chance of reinforcing appropriate behavior (Sulzer-Azaroff and Mayer 1977).

Physical Guidance

Another procedure used in the above program is *physical guidance.* With this technique the instructor physically assists the client through the appropriate behavior (e.g., taking the client's hand and putting it around the spoon and then scooping the food). Of course, the client must be relaxed and cooperative for

Figure 5.1. It is important for clients to develop independence by learning to eat without guidance. Staff stands behind John and places a spoon in John's hand.

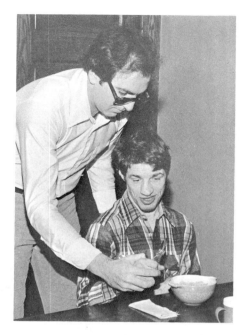

Figure 5.2A Shaping and physical guidance are effective techniques in training self-feeding behaviors.

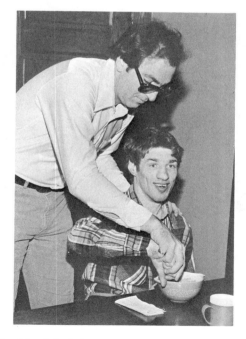

Figure 5.2B Staff guides John's hand and spoon into food.

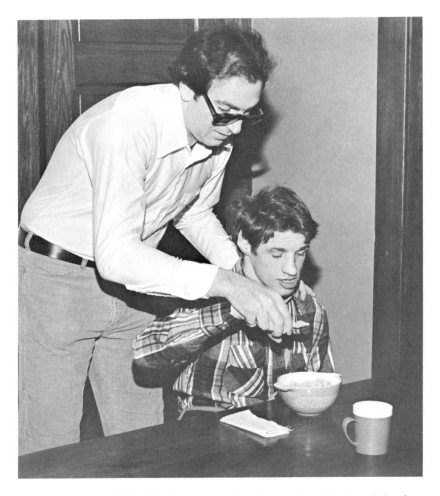

Figure 5.3. Staff lifts John's spoon to John's mouth and praises John for eating with a spoon.

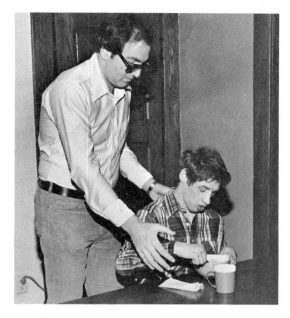

Figures 5.4. As John is able to do different parts of the eating program, the staff member should fade out physical guidance.

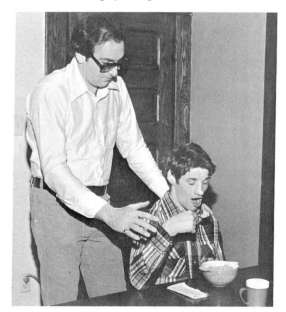

Figure 5.4B

Figures 5.5. As the client progresses, the staff member should stand back and allow the client to do it all on his own. Staff should be available to assist the client if necessary.

Figure 5.5B

Figure 5.6. SUCCESS.

this program to work. The instructor's assistance should be decreased as soon as the client is able to perform the task independently. Physical guidance should always be combined with verbal prompts or instructions ("John, pick up your spoon"). In this way verbal instructions become sufficient to prompt the client to perform the desired behavior without being guided physically. These instructions are important, since we want the client to succeed, and we may occasionally have to remind the client when to respond.

Modeling

Another technique that can be used in a training program is *modeling,* or demonstrating the appropriate behavior to the client. This technique is often used in toothbrushing, where the instructor goes through each step of the program and has the client imitate the process. Modeling should also be combined with verbal instructions. These instructions can cue the resident as to how to respond, so the instructor will not have to demonstrate the entire procedure each time.

67

Backward Training Procedure

Sometimes, new behaviors are best taught in reverse order, or *backward training procedure*. Dressing skills, for example, can be taught effectively through backward training used with physical guidance techniques.

PUTTING PANTS ON

Goal: *John will put on his pants without any help.*

Objectives	Procedures
1. John will pull his pants up from mid-thigh.	1. a) Place pants on John and pull up to his mid-thigh. b) Say, "John, pull up your pants." c) If no response, take John's hands, place them on his pants, and guide him through the motion of pulling up his pants. d) *Reinforce.* e) Fade out your assistance until John can pull his pants up from mid-thigh (use demonstration and verbal instructions).
2. John will pull his pants up from his knees.	2. a) Place pants on John and pull up to his knees. b) Say, "John, pull up your pants." c) If no response, take John's hands, place them on his pants, and guide him through the motion of pulling up his pants. d) *Reinforce.* e) Fade out your assistance until John can pull his pants up from his knees (use demonstration and verbal instructions).
3. John will pull his pants up from his ankles.	3. a) Place pants on John and pull up to his ankles. b) Say, "John, pull up your pants."

c) If no response, take John's hands, place them on his pants, and guide him through the motion of pulling up his pants.

d) *Reinforce.*

e) Phase out your assistance until John can pull his pants up from his ankles (use demonstration and verbal instructions).

4. John will put leg through pant leg and pull them up.

4. a) Place pants on one of John's legs.

b) Say, "John, pull up your pants."

c) If no response, physically guide John's putting his leg through the pants leg.

d) *Reinforce.*

e) Phase out your assistance until John can put his leg in the pants and pull them all the way up.

5. John will pull his pants up without assistance.

5. a) Give John a pair of pants.

b) Say, "John, put on your pants."

c) Offer assistance with getting his feet into pants, and phase out this assistance quickly.

d) *Reinforce.*

Reinforcing Each Appropriate Behavior

When a client is learning a new task, reinforce him or her after each desired behavior. Even if the staff member has to provide physical or verbal prompting, the client must still be reinforced after each step. Of course, as the program continues, the instructor will expect more and more of the client before giving rewards. This process of reinforcing each appropriate response as soon as it occurs is called *continuous reinforcement.* If the client is learning to pull up pants, each step of the program is reinforced until the client can perform it without any mistakes. Continuous reinforcement is the only procedure to use when teaching a *new* behavior. If an instructor trying to teach a client a new behavior only reinforces the client occasionally when he performs that behavior, the client will probably never learn that new skill.

Figure 5.7. Being able to dress independently gives clients a feeling of great satisfaction. Staff member places pants on John and pulls them up to his mid-thigh.

Figures 5.8. Backward training can be used with physical guidance techniques to train many dressing skills.

Figure 5.8B Staff says, "John, pull up your pants." If no response, the instructor takes John's hands, places them on his pants, and guides him through the motion of pulling up his pants. Then the staff member reinforces John, fading out assistance until John can pull his pants up from mid-thigh.

71

Figure 5.8C

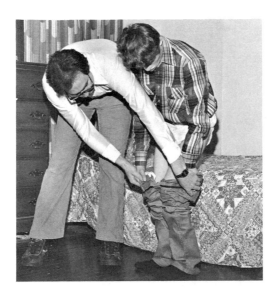

Figures 5.9. Staff places pants on John and pulls them up to his knees, fading out physical guidance when John can pull his pants up from his knees to his waist.

Figure 5.9B

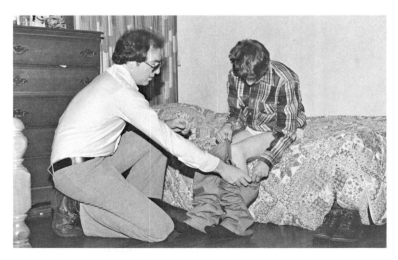

Figure 5.10. Staff places pants on John and pulls them up to his ankle. Then the instructor tells John to pull his pants up the rest of the way, using physical guidance if necessary.

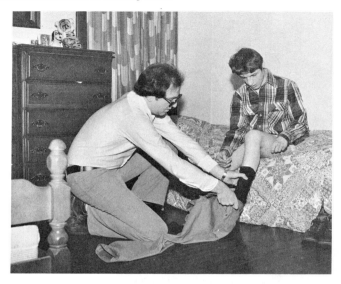

Figures 5.11. Staff places pants on one of John's legs and instructs him to pull them the rest of the way up.

Figure 5.11B John really feels good when he can do this much for himself.

Figure 5.12. Always remember to reinforce John whenever he gets his pants up to his waist.

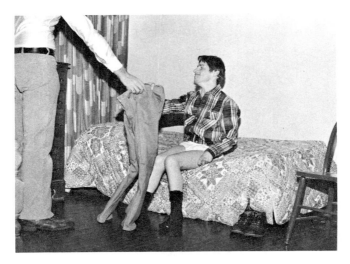

Figures 5.13. These pictures show John totally able to put on his own pants. A feeling of accomplishment is very satisfying.

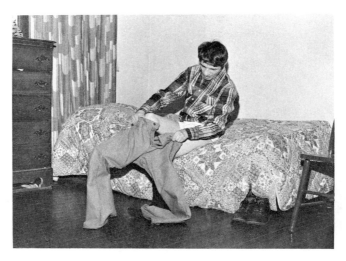

Figure 5.13B

Figure 5.13C

Figure 5.13D

Figure 5.13E

When learning a new skill, the client needs continuous feedback for performing the appropriate behavior. The reinforcement, however, does not always have to be the same. Here are some guidelines for using two types of rewards (food and praise) as continuous reinforcements:

1. Reinforce with food and praise after the client completes the step being taught.
2. Reinforce steps that the client has already learned with verbal praise as they are completed.
3. Reinforce the completion of the entire task with food and praise. Note that the client is always being reinforced for successfully completing a step, although the type of reinforcement may differ (Golden and Ho 1973).

Maintaining Desired Behavior

Once a behavior is learned and performed regularly, it should be reinforced less frequently. Now the resident should have to perform the desired behavior several times before being reinforced. This is called *intermittent reinforcement*. Intermittent reinforcement is used to maintain the behavior, or keep it going over time, after it has been learned through the use of continuous reinforcement. The transition from continuous to intermittent reinforcement should be gradual, or the behavior will begin to fall apart. If this occurs, return to continuous reinforcement until the behavior occurs on a regular basis, and then, more gradually than before, switch to intermittent reinforcement. Thus, once a client has learned to put pants on, the instructor would not reinforce each step, but would praise the client for wearing pants. When this becomes a regular event, the instructor would then reinforce the client every second or third time he or she saw the client get dressed unassisted. Finally, the instructor would praise the client only once in a while for this behavior.

ACTIVITIES FOR THOUGHT AND DISCUSSION

In class, break up into small groups. Each group will be assigned one of the following skills to teach a client, using either a shaping or backward training procedure:

    a) Putting on a shirt
    b) Brushing teeth
    c) Working a five-piece puzzle
    d) Drinking from a glass

Do each of the following:

1. Write a complete task analysis, including the procedures used to accomplish each goal.

2. Role-play the entire procedure (with one student playing the instructor and another the resident). Be sure to include the following in your role-playing:

    a) Types and amounts of reinforcers
    b) Instructions, physical prompting, and modeling, as necessary

3. All participants should receive feedback on their training techniques. Participants should practice each training procedure until the whole process flows smoothly.

Behavior Modification: Reducing the Frequency of Undesirable Behaviors 6

POSITIVE APPROACHES

There are many different ways to reduce or eliminate inappropriate behaviors. Several of these techniques involve the use of *positive reinforcement* and are, therefore, regarded as positive approaches to reducing the frequency of undesirable or dangerous behaviors. When using such an approach, we teach the client more appropriate ways of behaving as alternatives to the undesirable behaviors. Thus, we are not just getting rid of an inappropriate activity; we are also replacing it with an acceptable behavior. For example, if a client runs up and hugs and grabs other people whenever seeing them, the first thing that should be tried is to teach the client how to shake hands and say "hello."

We will always try to assist the client in developing new skills, rather than just focusing on getting rid of the troublesome behavior. This is an important consideration. We must assist the client in personal growth rather than limiting it. In a practical sense, this approach is more to the client's advantage. It helps the client learn new behaviors and gain self-esteem.

At times, the major focus will be on reducing an undesirable or unacceptable behavior. This is especially true when the client is dangerous to himself, to others, or to property. When a decision is made to reduce a dangerous behavior, positive approaches of training more acceptable behaviors should be tried first. Only when these do not work, should we try more negative procedures.

Limiting the Frequency of a Behavior

There are some behaviors that are annoying only because they occur so often. If they occured less often, they would not be as troublesome. *For these behaviors, the goal would be to have the behavior occur less often, rather than to eliminate it totally.*

For example, in a group home John bosses everyone around. We do not want to eliminate this behavior totally, but would like to get it down to bearable limits. How do we do this? After having John's approval that this is something he wants to work on, we collect data (using the observational techniques discussed earlier) to determine how often he bosses others on a daily basis. Let us say that it happens twenty-five times from 5:00 P.M. to 10:00 P.M., or about five times an hour. Then we set up a program to reinforce John for each hour that his bossing of others occurs four times or less. When this objective is met, we continue to reduce the maximum number of bossing statements until they are within bearable limits. Once John gets his bossing down to four times or less an hour, we have him work toward three times or less, then two times, and then once an hour. Finally, we work on reinforcing John for bossing others only once an evening. In this way John's inappropriate behavior is gradually reduced. He feels good about it, and so do the staff, because everything is done in a very positive manner.

This approach communicates to the client that a behavior is acceptable if kept within certain limits, and allows for a gradual reduction in the number of responses. This approach also has some disadvantages, however, which we must take into consideration when using it. It is time-consuming for staff, as it is a very gradual process. This procedure also focuses on the client's negative behavior, and we thus might accidentally reinforce this undesired behavior by putting much of our attention on it (Sulzer-Azaroff and Mayer 1977).

Reinforcing All Other Behaviors

Another positive approach to reducing undesired behaviors is to reinforce *all other* behavior, when the client does not perform the undesired act within a specified period of time.

Let me give you an example. On a ward in a state institution, one client would frequently bite the other clients' arms. First, we collected data on the frequency of his biting and the length of time he would go without biting. We found out that the client bit someone else between twenty and thirty times a day, but could go for up to twenty minutes without biting. Also, we knew that food was an effective reinforcer for this client. Then we started programming in such a way that the client could succeed. Since he could go for twenty minutes without biting, we broke his day down into twenty-minute intervals:

	20 minutes	20 minutes	20 minutes
8:00			
9:00			
10:00			
11:00			

The staff would set a timer for twenty minutes. When the bell went off, if the client had not bitten anyone during that interval, he would be reinforced with food and praise ("Good John, no biting!"). If he did bite someone before the end of the interval, he would immediately be placed in a chair away from everyone else for two minutes. The staff would mark the appropriate box on the chart, indicating that the client had bitten someone. When the timer went off after that interval, the client would not be reinforced; the timer would simply be reset and the program would start again. The technique just described also uses a negative method (time-out—removing John from the situation by having him sit in a chair for two minutes). This procedure will be described more fully later in the chapter.

When the client consistently (95 percent of the time) went without biting anyone for twenty-minute intervals, the interval was extended. Now John had to go for thirty minutes without biting before being reinforced. When he was able to do this consistently (95 percent of the time) we extended the time to forty-five minutes, then sixty, then ninety minutes, then three hours, five hours, eight hours, and so forth. Within three weeks, the client was no longer biting anyone.

This procedure has several advantages. It is a rapid technique that has long-lasting results. Above all, it is a very positive approach to changing behavior. Remember, when using this procedure, keep the time interval short at first, allowing the client to earn reinforcement often. Then gradually increase the interval as the client progresses (Sulzer-Azaroff and Mayer 1977).

This technique does have one major disadvantage, however, which we should take into consideration before deciding to use it. Since we are reinforcing any behavior except the undesired one, we might, by chance, reinforce a behavior that is worse. For example, Pearl is given a raisin every five minutes if she is not wringing her hands. But it just so happens that she is rocking at the end of five minutes, when the fruit is given to her. Her rocking behavior might increase as a result.

Reinforcing Competing Behaviors

Another positive approach to reducing undesirable behavior is to reinforce specific behaviors that compete with the unwanted activity. If Pearl is wandering around the workshop bothering people, for example, rather than punishing this behavior, we might reinforce her staying at the work station. If she spends more time at the work station, she will spend less time bothering others. A major advantage of this procedure is that results are long-lasting if the competing behavior is reinforced occasionally.

Before we could carry this program out, we would have to find out how long the client stays at her work area. Let us say that Pearl usually stays at her work table no longer than ten minutes before she wanders off. Staff should then reinforce her for every ten minutes that she is working (going over and checking her work, praising her for working, etc.). As the client consistently works for more and more ten-minute time periods, we can gradually extend the length of time that she has to work before being reinforced (twenty minutes, half an hour, one hour, two hours, full day). Of course, before carrying out this or any program, the client should be asked if she wants to change a specific behavior. We can only carry out a program if we have the client's approval.

ACTIVITIES FOR THOUGHT AND DISCUSSION

Develop programs to reduce the following behaviors using positive approaches. Review your programs in class.

- A. A client who slaps her head ten times an hour.
- B. A client who stares out of the window for a minimum of thirty minutes out of every hour.
- C. A client who interrupts other people's conversations at least three times an hour.
- D. A client who tears his shirt at least twice a day.

Be sure to think about what you want the client to be doing instead of these undesirable behaviors.

OTHER APPROACHES TO REDUCING UNDESIRABLE BEHAVIOR

Extinction   Several other methods can be used to reduce inappropriate behavior. One is simply to ignore the behavior. If Sam throws a tantrum when he does not get his way, for example, the staff can help Sam throw fewer tantrums by ignoring them when they occur. In behavior modification terms, this is called *extinction*—the withholding of all reinforcers when a client is doing an undesired behavior. This last statement is the key to using extinction effectively: *all* forms of reinforcement must be identified and withheld. This includes other people's reinforcing the client with attention (attention in almost any form, including being scolded, can be reinforcing). If the client receives reinforcement from any source, the behavior will not be reduced and in fact will probably occur more often. Using our example of Sam and his tantrums, if all staff agree to ignore Sam while he has a tantrum, but another resident runs up to help Sam when he screams, the program will be doomed to failure. Sam's tantrum is getting the result he wants—attention from others.

Besides not reinforcing the individual for inappropriate behavior, it is equally important to reinforce appropriate behavior. We want the client to act appropriately more often, and inappropriately less often, so the desired behaviors have to be reinforced.

If carried out effectively, extinction is a very long-lasting program. Staff, however, should be cautious when using extinction, because the undesired behavior usually gets worse before it gets better. Thus, staff should use this technique on a behavior that they feel they can endure if it takes a temporary turn for the worse (Sulzer-Azaroff and Mayer 1977).

Time-out from Reinforcement Another fairly common procedure, especially for parents, is to send the individual to sit in the corner or other quiet area for misbehaving. When this is done for a predetermined length of time whenever a specific behavior occurs, it is called *time-out*. In a time-out procedure, we remove the individual from reinforcing surroundings for a *brief* period of time (two to five minutes for severely or profoundly retarded individuals). The client should be taken to an area where he or she cannot be a part of the ongoing activities (a chair in the corner, a table on the far side of the room, the hallway). *Time out does not involve locking a person in a room alone.* When calm, the client should be removed from the time-out area.

For time-out to be effective, it must be applied every time the undesired behavior occurs. And, as stated previously, the time-out period should be kept short (Sulzer-Azaroff and Mayer 1977). This provides the client with an opportunity to get back into ongoing activities and act more appropriately. Also, the client should be reinforced after displaying appropriate behavior. Thus, we do not use time-out or extinction alone, but always combine them with other procedures that reward the client for acceptable responses. Time-out is a rapid technique with long-lasting results. At times, however, people get upset when placed in the time-out area, and caution should be used so that aggressive or self-abusive behavior does not increase.

Overcorrection Let us look at another one of Sam's annoying behaviors: When he gets upset, he turns over furniture. One technique that staff can use to reduce the frequency of this behavior is called *overcorrection*. This would involve, for example, requiring Sam to restore the room to a state much better than before he got upset (Sulzer-Azaroff and Mayer 1977). When Sam turns over the furniture, he should be expected not only to set it up straight, but also to dust and clean the room. This is usually an effective procedure if carried out consistently and immediately by the staff. There should not be a long delay between the act (upsetting the furniture) and the client's overcorrection of the act (cleaning up the room). The client *should not* be punished for a destructive act in addition to the overcorrection program. Also, the overcorrection should be related to the client's destructive act. Thus, a client who marks all over the walls should not be expected to do everyone's dinner dishes, but should have to wash the wall. In addition, the next time the client is near the wall with a magic marker, but writes on a piece of paper instead of the wall, he or she should be praised. Here too we should reinforce appropriate behavior.

One of the biggest drawbacks to overcorrection is that it requires a one-to-one ratio between staff and client. Thus, it is quite time-consuming for staff, and the close relationship between staff and client might end up reinforcing the client's undesirable behavior. Finally, overcorrection is a rather harsh technique and should be used with caution.

QUESTIONS

6-1. Look through some books on behavior modification (Sulzer-Azaroff and Mayer 1977 is a good one) for other techniques to reduce the frequency of behaviors. What are some of them, and what are some of the potential hazards of these techniques?

6-2. Compare extinction, time-out, and overcorrection. What are the advantages and disadvantages of each?

6-3. Describe some possible behaviors for which you would use each of the following: time-out, overcorrection, and extinction.

Training Self-Help and Daily Life Skills

7

With all the knowledge you have acquired thus far, you should now be able to train developmentally disabled individuals in skill areas to help them become more self-sufficient. To assist you with this process, some typical self-help programs have been included in this text. Programs in other areas (vocational training, socialization, and recreation) can be developed by using the same behavior modification techniques. The following programs should be practiced in a supervised role-play situation until the entire procedure, including reinforcing the client, flows smoothly and comfortably. After the programs have been practiced a sufficient number of times, work with your clients on training these skills.

The first step in any training program is to find out how much of the program the client can already perform. To do this, simply ask the client to complete the tasks without providing any instruction, prompting, or reinforcement. For example, in the hand-washing program, the staff would escort the client to the sink and say, "Wash your hands." Staff would observe and record what the client did. If the client could not complete the task, the staff would then instruct the client to perform each step of the task analysis (e.g., "John, turn on the water," "John, put your hands in the water," etc.), again observing and recording exactly what the client did on each step. After the staff knows what the client can do on each step, they can start training the steps in the order indicated. It is important to allow the client to continue doing any part(s) of the task that he or she already knows. For example, a client who can turn the water on and off but cannot wash should be allowed to turn the water on at the beginning of the program and off at the end, while learning all the other steps of hand-washing in between.

Each of the training programs to be discussed requires that staff devote individual attention to the client during the training period. To be effective, these programs must be carried out consistently and on all shifts. Thus, if a client is being taught to eat with a spoon, the program should be followed at every meal. To achieve this one-to-one interaction, it is often necessary to attract employees or volunteers from other areas to assist with the training. If this is not possible, the training should be done with only a few clients, and then as they acquire the skills, additional clients should be introduced to the training process.

Occasionally, a client may progress well through the program, and then suddenly stop learning or appear to have forgotten what has already been learned. When this occurs, the instructor should find the last step of the task analysis that the client can complete satisfactorily, and start training again from that last step. This procedure may have to be repeated several times (Watson 1972). If progress with a client continues to be slow, consult the behavior modification specialist or the psychologist for assistance. Several possible causes for the lack of progress might be examined, including the strength of the reinforcer (Might there be something more powerful to motivate the client?) and the complexity of the step being trained (Does the program have to be broken down into smaller steps?).

TOILET TRAINING

The week before training is to begin, the staff should collect data on what times the client defecates or urinates, and whether or not toilet facilities are used. The easiest way to do this is to break the client's waking day into thirty-minute time periods and mark a chart with the client's toileting data, as shown in Exhibit 7.1.

Once this data is collected, a toileting schedule can be planned around the times that it would be most likely for the client to eliminate. From the example in Exhibit 7.1, the client would be taken to the bathroom upon waking, about 9:00 A.M. (to defecate), and about 10:30 A.M., 1:00 P.M., 4:00 P.M., 7:30 P.M., and 10:00 P.M. (to urinate).

The client should be escorted to the bathroom at the earlist times that he or she is most likely to use it. Thus, a client who pretty consistently urinates between 10:30 and 11:30 should be taken to the bathroom at 10:30 and if necessary, at ten-minute intervals after that, until he or she eliminates. After the toileting schedule is established, use the following procedure to teach the client to eliminate on the toilet. Other programs can then be used to teach the client to 1) go to the bathroom without assistance, 2) to pull pants down and up, 3) to wipe himself or herself, and 4) to flush the toilet. (This program is based on Watson 1972.)

	7:00	7:30	8:00	8:30	9:00	9:30	10:00	10:30	11:00	11:30	12:00	12:30	1:00	1:30	2:00	2:30
Monday		(U)			(D)				(U)					(U)		
Tuesday	(U)					(D)			(U)							(U)
Wednesday	(U)			(D)				(U)						(U)		
Thursday	(U)					(D)		(U)					(U)			
Friday		(U)			(D)			(U)					(U)			
Saturday	(U)				(D)				(U)				(U)			
Sunday		(U)				(D)			(U)					(U)		

	3:00	3:30	4:00	4:30	5:00	5:30	6:00	6:30	7:00	7:30	8:00	8:30	9:00	9:30	10:00	10:30
Monday			(U)							(U)					(U)	
Tuesday				(U)							(U)					(U)
Wednesday				(U)							(U)				(U)	
Thursday			(U)								(U)				(U)	
Friday				(U)						(U)						(U)
Saturday			(U)								(U)					(U)
Sunday			(U)							(U)						(U)

U = Urinates on toilet

D = Defecates on toilet

(U) = Urinates on self or elsewhere

(D) = Defecates on self or elsewhere

Exhibit 7.1. Toileting Data.

TOILET TRAINING

Goal: *Sam will eliminate while seated on the toilet.*

Objectives	Procedures
1. Sam will go to the bathroom when told.	1. Say, "Sam, go to the toilet." If no response, a) escort Sam to bathroom; b) pull down his pants (if Sam knows how to pull down his pants, ask him to do so).
2. Sam will sit on the toilet when told.	2. Say, "Sam, sit down. (If no response, this step should be taught separately, and when Sam can sit on the toilet for up to five minutes, then start the toilet training program.)
3. Sam will eliminate while seated on the toilet.	3. Encourage Sam to eliminate and *reinforce* him when he does.
4. Sam will stand up when told.	4. Say, "Sam, stand up." If no response, a) prompt him physically, and then fade back to just the verbal direction; b) pull up Sam's pants (if Sam knows how to pull his own pants up, have him do so by saying, "Sam, pull up your pants"); c) *reinforce* with praise; d) return Sam to activity.

A similar toileting program can be used for bedridden clients. Instead of taking these people to the toilet, a bedpan can be brought to them according to their own individual schedule.

SHOWERING

The following showering program should be taught one step at a time. Teach the first step, and then do all the rest for the client (except, of course, those steps that the client already knows). Reinforce at the end of the step that the client is being taught, as well as when the whole procedure is completed. As the client progresses, he or she should perform all the steps up to the one being taught. A separate program should be used to teach the client how to dry off after the shower. (This program is based on one developed by Watson 1972.)

SHOWERING

Goal: *Pearl will shower independently.*

Objectives

1. Pearl will soap the washcloth when given one.

2. Pearl will wash her whole arm and hand.

Procedures

1. Say, "Pearl, wash yourself." Give her a washcloth and soap and say, "Pearl, soap the wash-cloth."
   a) If no response, model the behavior, then guide Pearl's hands if there continues to be no response.
   b) *Reinforce.*
   c) Fade the physical guidance and modeling.
   d) When Pearl can soap the washcloth with just a verbal prompt, start teaching step 2.

2. Say, "Pearl, wash your arm."
   a) If no response, model the behavior on your body for Pearl to imitate.
   b) If still no response, use physical guidance.
   c) Fade the physical guidance and the modeling.
   d) *Reinforce.*
   e) When Pearl can wash her arm with just a verbal prompt, start teaching step 3.

3. Pearl will wash her other arm.

3. Say, "Pearl, wash your other arm."
   a) If no response, model the behavior on your body for Pearl to imitate.
   b) If still no response, use physical guidance.
   c) Fade the physical guidance and the modeling.
   d) *Reinforce.*
   e) When Pearl can wash her arm with just a verbal prompt, start teaching step 4.

4. Pearl will wash her face and neck.

4. Say, "Pearl, wash your face."
   a) If no response, model the behavior on your body for Pearl to imitate.
   b) If still no response, use physical guidance.
   c) Fade the physical guidance and the modeling.
   d) *Reinforce.*
   e) When Pearl can wash her face with just a verbal prompt, start teaching step 5.

5. Pearl will wash her leg and foot.

5. Say, "Pearl, wash your leg."
   a) If no response, model the behavior on your body for Pearl to imitate.
   b) If still no response, use physical guidance.
   c) Fade the physical guidance and the modeling.
   d) *Reinforce.*
   e) When Pearl can wash her leg with just a verbal prompt, start teaching step 6.

6. Pearl will wash her other leg and foot.

6. Say, "Pearl, wash your other leg."
   a) If no response, model the behavior on your body for Pearl to imitate.

b) If still no response, use physical guidance.

c) Fade the physical guidance and the modeling.

d) *Reinforce.*

e) When Pearl can wash her other leg and foot, start teaching step 7.

7. Pearl will wash her whole body without verbal prompts.

7. After Pearl has been trained on all the above steps, begin the next training session with "Pearl, wash youself." Fade out verbal prompts. Reinforce with praise each step successfully completed. Then fade out reinforcement until the whole task is successfully completed.

DRESSING PROGRAMS

Many dressing programs are taught in reverse order. That is, they start with the last part that the person normally does, and train that step. Then these programs train backwards until they get to the beginning of the procedure. Also, since undressing skills are usually learned before dressing skills, they should be taught first. Loose fitting clothes should be used in training as these are easier to put on and take off.

TAKING OFF PANTS

Goal: *John will take off his pants.*

Objectives

1. John will take pants off from his ankles.

Procedures

1. With pants pulled down to John's ankles, say, "John, take off your pants."

a) If no response, use modeling or gestures to assist the client.

b) If still no response, use physical prompt.

c) *Reinforce.*

d) Fade out physical prompts and modeling.

2. John will take pants off from his knees.

   e) When John can take his pants off when told, start training step 2.

2. With pants pulled down to John's knees, say, "John, take off your pants.

   a) If no response, use modeling or gestures to assist the client.
   b) If still no response, use physical prompts.
   c) *Reinforce* when pants are off.
   d) Fade out physical prompts and modeling.
   e) When John can take his pants off when told, start training step 3.

3. John will take pants off from his mid-thigh.

3. With pants pulled down to John's mid-thigh, say, "John, take off your pants."

   a) If no response, use modeling or gestures to assist the client.
   b) If still no response, use physical prompts.
   c) *Reinforce* when pants are off.
   d) Fade out physical prompts.
   e) When John can take his pants off when told, start training step 4.

4. John will take pants off from his waist.

4. With pants at John's waist, say, "John, take off your pants."

   a) If no response, use modeling or gestures to assist the client.
   b) If still no response, use physical prompts.
   c) *Reinforce* when pants are off.
   d) Fade out physical prompts and modeling.

(For a program to teach putting pants on, see pp. 68–69.)

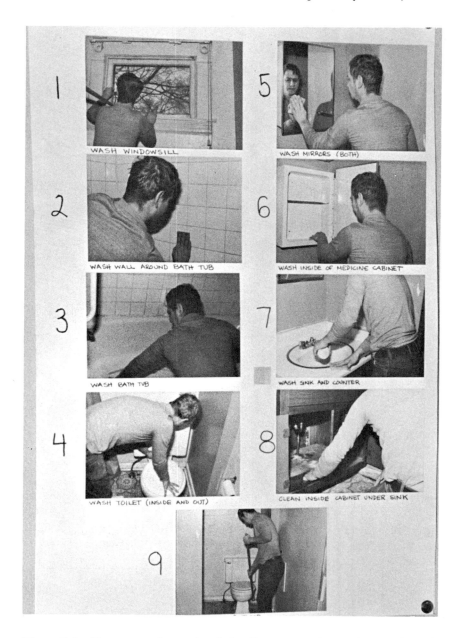

Figure 7.1. Pictures can be used to help clients remember each step of a task.

QUESTIONS

7-1. Write the procedures (what the staff will be doing) for each step of the following programs:

    a) Taking off socks

        1. Client will pull sock off from toe.

        2. Client will pull sock off from over heel.

        3. Client will pull sock off unassisted.

    b) Putting on shirt

        1. Client will pull shirt down from chest when head and arms are already through.

        2. Client will put one arm through and pull the shirt down if head and one arm are already through.

        3. Client will put both arms through and pull shirt down if head is already through.

        4. Client will put head and arms through and pull shirt down when given to client in the correct position.

        5. Client will put shirt on completely and unassisted.

7-2. Write the objectives (what the client will be doing) for the following programs:

    a) Washing face

    b) Putting on socks

    c) Washing hands

(Turn to appendix A for answers to question 7-2.)

ACTIVITY FOR THOUGHT AND DISCUSSION

View the film *Genesis*. This film demonstrates the training techniques just discussed and can be ordered from Educational Division, Hallmark Films, 51–53 New Plant Ct., Owings Mills, Md. 21117.

Other Client Needs   8

COGNITIVE DEVELOPMENT

Cognitive development is the development of one's intellect: the ability to
reason, to determine cause-and-effect relationships, and to gain control over the
environment. In short, cognitive development is the development of one's
thinking abilities. It is now known that to develop thinking skills, a person must
be as active as possible and must use all the senses available. Thus, clients should
be exposed to situations where they are touching, smelling, hearing, tasting, and
seeing a variety of objects and events.

An environment in which the surroundings and routine stay the same for
days on end is a boring environment. When placed in a situation that does not
change, many people simply "turn off" and do nothing for most of the day.
Others seek to make their surroundings more exciting by stimulating themselves (walking, falling, throwing things, masturbating) or by creating chaos for
others. In either case, people in boring environments do not learn good coping
skills, do not learn how to appropriately control their surroundings, and do not
develop to their full potential.

To assist you in understanding the need for experiences and activities
that promote thinking abilities, we shall review the normal course of cognitive
development. Try to relate this information to the behaviors that your clients
are displaying. (Much of the following material is based on a theory of cognitive
development proposed by Piaget).

Birth to Two Years of Age

Shortly after birth, newborn infants' behavior is mostly reflexive. Infants do not
think, but when stimulated respond automatically. Some of these early reflexes

include sucking, grasping, crying, and moving the arms. When anything is placed in newborns' mouths, they will suck it. Mostly through experience, infants begin to learn the difference between objects that produce food when sucked (bottle, mother's breast) and those that do not.

At birth infants know nothing about objects in the environment or about their own bodies; only through being active—through seeing, hearing, touching, smelling, and moving—do they learn about the world and themselves. At first, many of these actions are random and accidental, but if an accidental motion leads to something pleasant, the child will attempt to repeat that motion. For example, even a three-day-old infant can learn to control his or her surroundings. If a plastic nipple is attached to a mobile so that sucking the nipple causes the mobile to move, the infant can learn how to make the mobile move by sucking on that nipple. In this manner, infants learn different motions that have different effects on their bodies as well as different effects on their surroundings. This allows them to begin adding new ideas about these surroundings and to change some behaviors in an attempt to meet environmental demands. Through experience, children learn cause-and-effect relationships, and their behavior becomes more purposeful (Robinson and Robinson 1976).

Very young infants are most concerned about their own bodies and their own activities. As they become more aware of their bodies, however, their interests turn outward, towards the world around them. Eventually, as infants develop, they begin to relate events in their surroundings to themselves. Thus, if they see mother putting on her coat, they begin to cry because they realize that she is leaving.

As children continue to progress, they begin to search out new experiences, to explore the environment, and to experiment with things they find (Robinson and Robinson 1976). Children at this stage, observing mother turn on a light, will attempt to experiment with turning the light on and off and will receive much pleasure from this activity. During this stage, they also learn that objects are permanent and do not disappear when out of sight. The important thing to remember about this stage is that all that children know results from their own actions. Thus, to be able to do something, children must imitate behavior they have observed. Merely watching a new behavior, however, is not enough for children to learn: the very young know by doing. Just watching mother turn on the light is not enough for them to learn this act. To master it, children must do it themselves.

Cognitive Stages and the Developmentally Disabled

All the preceding events usually occur within the first two years of life. You will notice, however, that many severely and profoundly mentally retarded adults are functioning in similar ways. Some of the ways these people interact with the environment are random and lack intent, and sometimes, they do not recognize the effects of their behavior on their surroundings. For example, a client may accidentally bump the laundry cart and not respond fully when it tips over and crashes. For such clients, it would be helpful to learn that their actions do have an effect on their surroundings. One simple way to do this is to place a penny in an empty soda pop can so that clients can use it as a rattle; whenever they shake it, they will be causing some change in their surroundings (making a noise).

Other severely or profoundly retarded clients are somewhat more advanced. When they tip over the laundry cart, they enjoy the noise and commotion and attempt to do it again. In addition, they may begin to explore and try new things, seeing what sounds other objects make when they are tipped over, turning the lights on and off constantly, or getting into closets. Often staff see these behaviors as inappropriate and something to be stopped. These behaviors, however, can also be viewed as movement toward intellectual growth and should be dealt with accordingly. Clients who display them should be provided with equipment to help them explore, experiment, and thus learn about common objects. In learning to operate some of this equipment, they learn to control some parts of their environment. For example, many severely retarded people enjoy music and can learn how to turn a portable radio on and off. They can also be provided with an area such as a drawer, which houses a variety of simple objects, in which they can do as they please. When clients begin to explore an off-limits area, the staff can redirect them to the appropriate "exploration" area. Of course, new materials should constantly be introduced to this area to keep the clients' interest.

To assist clients in learning about the permanence of objects, staff can play a version of the children's game hide-and-seek. Objects that a client is familiar with can be hidden from view. At first, they can be hidden while the client is watching; later, as the client understands the idea of looking for the items, they can be hidden while the client is not looking.

Many of these ideas are recreational in nature. Such activities can be very helpful in teaching clients about the environment and thus in helping them learn about themselves and their surroundings.

QUESTIONS

If you are working with severely or profoundly retarded clients:

8-1. How can you help the clients learn more about themselves and the environment?

8-2. What can you do to help clients gain some more control over the environment?

Years Two to Seven

At approximately one year of age children normally begin using language. As they grow older and develop more verbal skills, they can also solve more problems. Language frees them from dealing with only what is happening at the moment, and allows them to think as well about the past and the future. In a world no longer dominated by physical objects, children's ideas and thoughts expand rapidly. Language allows them to begin applying what they observe around them to their own behavior. They no longer know only by doing, but can now use words instead of actions to learn new ideas and behaviors.

Children in this stage learn by watching others. If a child sees a friend pulling a dog's tail and the friend is then bitten by the dog, the child does not have to do the same thing to learn that one should not pull a dog's tail. Now he or she can learn simply by watching others or by being told about things ("If you pull that dog's tail, it might bite you"). An individual with language can also begin to share ideas with others and find out more about their feelings and attitudes. Thus, language also promotes socialization.

Although children in this stage (usually between the ages of two and seven) have taken some giant steps towards expanding their thinking abilities, many limitations still exist. For one thing, individuals in this stage of cognitive developement have difficulty seeing another person's point of view. A problem or event exists only from their point of view; so it is very difficult to explain to them the errors of their thinking, for even if logic is used, these individuals are generally unable to reason rationally.

During this period, games with rules are usually learned. These games are not learned quickly, however, as these individuals often make up their own rules as the game goes on. Only through interacting with others do people learn to accept standard rules and other viewpoints. For children to learn this, they need to see other people doing things differently than they are or talk to other children and adults about the events they are experiencing.

Many moderately mentally retarded individuals attain this level of cognitive development. These people have difficulty understanding social values and other people's points of view. They might be able to understand that a certain behavior is not acceptable, but probably could not provide reasons why it is wrong beyond the rules that they have memorized. They also may have difficulty accepting rules that conflict with their ideas of how things should be. The moderately mentally retarded can relate experiences, however, and can learn simply by observing others. They can plan activities and make decisions. They are capable of sharing their experiences with others. Only the interaction with others and their surroundings will allow these individuals to gradually know and understand the world around them to a greater degree.

QUESTIONS

If you are working with moderately mentally retarded clients, examine how you are interacting with them.

8-3. Are you providing them with ample opportunities to interact and learn from others?

8-4. Do you provide opportunities for your clients to make decisions and see the consequences of their choices?

8-5. Do you allow them to use the language skills that they have acquired?

Late Childhood

As children continue to develop, their thought processes become more complex and organized, and they are able to think more logically. They learn to assume other people's points of view, understand logical rules, and play games with complex rules. As they continue to develop, their thought processes become more abstract, and these children begin to make assumptions about events they see. These assumptions, based on logical processes, allow children's minds to roam far beyond the immediate situation. This capacity for abstract thought usually emerges in children over eleven years of age. Mentally retarded individuals generally do not attain this level of complex thought and generally respond to the world around them in a concrete fashion (thinking about things and events).

Summary

The important thing to remember about cognitive development is that we learn by doing (Robinson and Robinson 1976). This is especially true of mentally retarded individuals. When teaching them, it is often helpful to provide demonstrations or model the desired behavior, rather than lecturing clients about what they should be doing.

Another important way people learn is through interaction with others. This allows for an exchange of ideas and can assist the individual in understanding the viewpoints of other people.

Finally, it is important to remember that people do not change their behavior or learn new ideas in a short and sudden process. Rather, cognitive development occurs gradually (Robinson and Robinson 1976). Thus, mentally retarded individuals must be taught new concepts and behaviors in a step-by-step procedure, which makes certain they have mastered one step before teaching the next.

LANGUAGE AND COMMUNICATION

Importance of Language

As we discussed previously, language plays a very important role in an individual's development of reasoning abilities. Spoken and written language (English, French, German, etc.) are the most common forms of communication. Does this mean that people who have difficulty speaking cannot communicate? Of course not! There are many ways that people can improve their spoken language or learn to communicate through other means, such as gestures, sign language, or communication boards. Understanding that people can learn to communicate despite many handicapping conditions is especially important when working with mentally retarded and other developmentally disabled persons, because there appears to be a higher incidence of communication disorders among these groups (Keane 1972).

Regardless of the way we learn to communicate (speech, gesture, or communication board), communication is a basic need of *all* people, which is essential to interaction as well as self-sufficiency. It is especially important in teaching, for without the ability to communicate, ideas cannot be exchanged and people cannot learn (Keane 1972). The ability to communicate also allows people to express their feelings and needs. When people are unable to communicate effectively, they usually become frustrated and angry and begin to feel badly about themselves, because they cannot get their ideas across to others. This inability to communicate may also lead them to avoid others and perhaps even to withdraw from activities.

ACTIVITIES FOR THOUGHT AND DISCUSSION

A. Sit across from another student in class and try to find out the following information about that person, *without speaking or using gestures.* Try this for three minutes.

1) Name
2) Age
3) Address
4) Hobbies
5) Favorite color

Obviously, you cannot communicate too well if you cannot talk or use gestures. Were you frustrated? Were you getting angry? Did you want to stop the activity?

B. Now, try to find out the same information using only gestures and no speech. You have three minutes to do this. What information did you find out about the person across from you? Gestures do allow communication to occur.

C. Now find out more about the person by using speech. How much more information did you acquire? Were you as frustrated as when you were not allowed to speak?

Language Development

Early Forms of Communication Communication between mother and child begins within the first few moments after birth. Newborns are able to communicate with their mothers when uncomfortable or hungry, usually by crying. Within a few months, babies begin to use different cries to tell parents what the problem is. One tone may indicate hunger, another boredom, and yet another that diapers need to be changed. During the next few months, infants learn to express other emotions besides discomfort. Learning to gurgle, laugh, and coo, they can now express pleasure and interest. These new skills move children closer to talking (Murphy and Leeper 1974).

First Words Children's first words usually occur at about one year of age, when they make some sound and it brings a very positive response from the parents. Saying "Ma-Ma" or "Da-Da" is rewarded with hugs and smiles. Thus children tend to repeat these words and also begin experimenting with sounds to see which ones provide a response. Saying "Wa-Wa" might bring a glass of water, for example. Or they might try to imitate words that others say to them. Thus, a mother might say, "Would you like a cookie?" and her child might respond with "Ka-Ka." Mother and child are now beginning to communicate with words (Murphy and Leeper 1974).

Learning to use words effectively takes a lot of practice and time. It also requires a lot of encouragement from others. This is especially true for developmentally disabled children and adults who are learning new ways to communicate their needs, feelings, and thoughts.

Types of Speech Problems Many developmentally disabled individuals have speech difficulties. These may take several different forms (Bensberg 1965).

1. *Delayed speech* is very common among developmentally disabled children. It means, simply, starting to talk later than normal.

2. *Articulation defects* are an inability to make speech sounds properly, perhaps because of physical and muscular problems, as found in individuals with cerebral palsy.

3. *Stuttering* occurs when the rhythm of speech is disturbed.

4. *Problems with understanding speech* do not always mean that a person has a hearing problem. Rather, because of some learning difficulty, the individual does not put meaning to words that are heard.

5. *Inability to speak* does not necessarily mean that a person is unable to communicate, only that spoken words are not used to express thoughts.

Helping Residents to Communicate

Talk to Clients   How can we help residents who display some of these communication problems? One thing to remember is that people can usually understand many words before they can say them. (This may not be true, of course, for people who have difficulty understanding speech, as described previously). Therefore, it is very important to talk to your clients, even if they cannot speak to you. You can talk to your clients during many of the activities that they participate in. For example, objects and actions can be labeled as clients get dressed (shirt, pants, putting on socks, zipping), and food can be named as they eat (turkey, bread, milk). Even if a client does not understand all that is being said, the idea that he or she is an individual worth your attention is still being communicated. Also, when clients begin to associate words with objects and activities, they will be able to understand what is going on around them more fully.

Consult the Language Development Specialist   Other techniques can be used with clients who have severe communication problems. These techniques and programs should be developed more fully by a *language development specialist* and should not be implemented without complete directions from this professional. (These specialists may have different titles in different facilities and may be called *speech therapists* or *speech and language pathologists*.) The language development specialist knows how to assess clients with communication problems and should be consulted for assistance in developing appropriate programs for such clients. The ideas presented here are merely to provide you with some thoughts about what can be done to help clients communicate more effectively.

Use Gestures and Sign Language   One method of communication that does not depend on speech is the use of gestures or sign language. *Signing* (using sign language) is helpful for people who have difficulty understanding speech or have problems with speaking. People who use signing, however, must have good control of their hands and fingers, the tools they will use to communicate. The simplest form of *gesturing* is to point to the desired object. A client who wants a drink of water, for example, can point to the sink. Signing is more complex than gesturing in that different symbols of events and objects can be made by changing the position of the hands. The object need not be present for the person to communicate, and this allows much more freedom of expression. The client, however, must be able to remember the different signs and how to form them with the hands.

Figure 8.1A   This person is using sign language to say "Hello."

Figure 8.1B

Figure 8.1C

Figure 8.1D

Figure 8.1E

Figure 8.2. These are some examples of pictures that can be used on a communication board.

Figure 8.3. Pictures can be used to communicate ideas when people have difficulty reading.

Try Communication Boards Many developmentally disabled individuals have multiple handicaps (more than one disability) and lack good motor control. For these people, using sign language would be difficult. They can better communicate by using *communication boards*. These boards can take many forms and should be geared to the ability level of the client.

In the simplest form a series of pictures shows the client's basic needs (toilet, glass of water, food). The staff point to one picture at a time, and the client gives some prearranged signal to indicate either "yes" or "no." This simple procedure can be used regardless of the severity of the handicap. All that is necessary is that the client have some movement to use as a signal to others (Vanderheiden and Harris-Vanderheiden 1976).

Clients with somewhat better cognitive skills can use communication boards with words, symbols, or the alphabet printed on them. If these clients also have better muscle control, they can point to the words or letters and communicate more complex ideas. In these ways they can have conversations with others, rather than just communicating their basic needs.

INTERACTING WITH THE CLIENT

As discussed previously, communication is an important skill and can take many forms. Whatever form it takes, communication with clients should always be on a person-to-person basis. Staff should not belittle clients in any way, but should make every effort to understand what they are trying to communicate. Clients' needs and desires should be honored as much as possible and their questions answered without sarcasm or anger. In this way staff will communicate that they respect each client as a person, and this will enhance the client's self-respect.

Other methods of interaction also promote a good relationship between staff and clients. There are many nonverbal ways of communicating acceptance, such as smiles, pats on the back, and hugs. In addition, doing something special for clients also communicates that they are special to you.

Emotional Needs

All these interactions are important to clients in helping them meet their emotional needs. Like the rest of us, clients under our care need love, security, and respect. They need to know that there are people who can be depended on to assist them in meeting these needs. They need to be informed about what is happening around them, and to them. They need to know what is expected of them, and they need to be able to express what they expect of you.

Client Frustrations

Despite having the same emotional needs as the rest of us, many clients become frustrated because of the difficulty they have in meeting these needs. Some of these frustrations are due to their handicaps, which limit people's chances for success and create more failure experiences. Clients with communication disorders may become frustrated when others do not understand them well. Clients with mental retardation may be frustrated if they cannot achieve all that they hope to achieve (driving a car, marriage, becoming rich). When people fail at things that they are trying to achieve, they often become easily irritated and may have a tendency to give up easily. After repeated failures, people tend to feel badly about themselves and may not be motivated to be a part of ongoing activities.

Staff may be helpful to clients by being supportive: encouraging them to participate in activities and providing praise when they succeed. The staff can also be helpful by providing experiences for clients that insure success. Recreation, leisure time, and play activities are often successful activities for clients, allowing them to release much of their tension and to learn more about themselves.

QUESTIONS

8-6. How do you help your clients meet their emotional needs of security, love, and respect?

8-7. What do you do to communicate your acceptance of your clients? How do you do this verbally? Nonverbally?

8-8. In what ways do you treat your client as less than equal to you? How can you change these habits to develop more of a person-to-person interaction?

Recreation and Leisure Time

9

By now, you have seen that residents should be involved in activities so they can learn about themselves as well as their surroundings. These recreational activities should be a planned portion of each day. Recreational and leisure-time experiences are activities that a person does for pleasure, which provide a release from the routine and refresh both mind and body. These activities, however, are also important for clients in improving their sensorimotor, cognitive, communication, and socialization skills (Luckey and Shapiro 1974).

FUNCTION OF RECREATION

The Ability to Play and Interact with Others Changes over Time

Changes in one's recreational skills and interests occur as one grows older and one's thinking abilities change (cognitive development). Your clients' recreational skills and interests also change with age and experience. In order to help them develop further, you need to understand the skill level at which they are functioning so you can plan appropriate activities to encourage continued development. Remember, all people are capable of learning, regardless of their disability or age.

The basis for recreational activities lies in *play*. As individuals establish control of their bodies by being able to sit, balance, and coordinate hand motions, play changes from random motions to repetitive activities. Emptying and refilling a box of blocks is a favorite activity for people at this developmental level. These people can now play by themselves for longer periods of time. Play at this level helps individuals learn ideas about the size, position, and texture of objects.

As children continue to develop, toys and games are used more in their play activities. During this later period children first learn to interact and cooperate with each other, making use of imagination and creativity. As they develop further, games with rules are introduced, which make them more aware of social demands. People become more interested in clubs and hobbies, as well as heterosexual (males and females doing things together) activities such as dances and parties (Luckey and Shapiro 1974).

For this normal progression of changing skills and interests to occur, space and equipment must be made available. Most importantly, recreation should be a planned part of each resident's daily program.

At times clients may have to be taught how to take part in specific activities. Other clients may already know how to engage in the activity, but need to be prompted to participate. Still other clients, who are willing to take part, may have difficulty choosing between activities. Each client should be worked with according to his or her particular needs and level of ability.

Recreation for the Profoundly Retarded

It has been shown that even profoundly retarded adults can be taught recreational and social activities when appropriate materials are made available and clients are encouraged and taught how to use them. For profoundly mentally retarded individuals, preschool activities and supplies are often effective, including fingerpaints, tambourines, rhythm sticks, Play Doh, blocks, and tinkertoys. If the appropriate activities are planned, staff will usually find that the clients get more out of the activity than just a good time. Some important side-effects of recreational activities for clients are improved attention span, fewer problem behavior, better ability to follow verbal directions, and improved staff morale (Quilitch and Grey 1974).

Other Benefits of Recreation

We can see that although the goal of recreational activities is usually enjoyment or fun, these activities also provide a chance for clients to learn about their environments, gain control over their bodies, release built-up energies and frustrations, and develop self-esteem. They learn appropriate ways to stimulate themselves and thus are less bored and less likely to become aggressive toward themselves or others. Once people learn to enjoy themselves and have fun, they are more likely to be motivated to participate in other activities as well. Not surprisingly, clients often become more cooperative once they regularly take part in recreational programs.

Figure 9.1. Learning leisure-time skills is very important.

GUIDELINES FOR INTERACTING WITH CLIENTS

1. *For clients with extremely short attention spans, it is best to plan activities that last for only a few minutes.* Then, present these activities to the client frequently (every half-hour or so) throughout the day.

2. *Some clients need to be worked with individually.* This is especially true if you are teaching them a new skill. Working in small groups (two or three people), however, is often effective as well, especially if you are trying to build socialization skills.

3. *Remember your behavior modification techniques.* You may need them to teach clients some recreational skills. If you need to, review the earlier sections in this book on task analysis, reinforcement, and shaping. Below is a sample program that can be used, step by step, to teach a client how to put together a four-piece puzzle.

   a. The client will put one piece into the puzzle when it is placed next to the appropriate space and when the three other pieces are in the puzzle.
   b. The client will put one piece into the puzzle when it is placed outside the puzzle frame and when the three other pieces are in the puzzle.
   c. The client will put two pieces into the puzzle when they are placed next to the appropriate spaces and when the two other pieces are in the puzzle.

    d. The client will put two pieces into the puzzle when they are placed outside the puzzle frame, but near the appropriate spaces, and when the two other pieces are in the puzzle.

    e. The client will put two pieces into the puzzle when they are placed outside the puzzle frame and when two other pieces are in the puzzle.

    f. The client will put three pieces into the puzzle when they are placed next to the appropriate spaces and when one other piece is in puzzle.

    g. The client will put three pieces into the puzzle when they are placed outside the puzzle frame, but near the appropriate spaces, and when one other piece is in the puzzle.

    h. The client will put three pieces into the puzzle when they are placed outside the puzzle frame in random order and when one other piece is in the puzzle.

    i. The client will put four pieces into the puzzle when they are placed next to the appropriate spaces.

    j. The client will put four pieces into the puzzle when they are placed outside the puzzle frame, but near the appropriate spaces.

    k. The client will put four pieces into the puzzle when they are placed outside the puzzle frame and in random order.

    l. The client will dump four pieces out of the puzzle and put them back in.
(Golden and Ho 1973)

4. *Progress is often slow.* Learn to expect small changes in the client's behavior. If you are using a task analysis procedure, recognize that the client's learning a new step or a portion of a new step is a major accomplishment. Keep providing encouragement. *Remember, the way that you respond to the client will often influence his or her motivation.* Your attitudes are important.

5. *The activity should be fun, as well as a learning experience.* Thus, if an activity is not going well, make the activity more suited to the client's needs or change to another activity. Find the right activity and the right reinforcer so that the client will want to participate.

6. *Activities should not be limited to what can happen indoors.* Use the outdoors for activities as well. Let clients experience the change of seasons (e.g., collecting leaves in the fall, making snowballs in the winter), new settings, and new people. Some inexpensive and fun activities that most people enjoy include

Going for a walk

Playing table tennis

Going to the movies

Talking

Going shopping

Calling a friend on the phone

Going out to eat

Hobbies (gardening, needlepoint, and so on)

Going to the fair or the circus

Visiting friends and family

Going to a concert

Having a picnic or other outing

Going swimming

Bowling

Watching a ball game

Playing baseball or basketball

Riding a bike

7. *Different people enjoy different activities.* When planning a recreation program, think about your client's special interests and include them in the program.

QUESTIONS

9-1. List all the activities you think would be appropriate for your clients. Be creative. Think not only of what can be done at your facility, but also of things that can be done in the community.

9-2. From this list of activities, plan a recreation schedule for your clients for one week. Include individual and small group activities.

9-3. Write a program that you can use to teach a client how to stack five rings. Try to teach your program to one of your clients. (*Turn to appendix A for a sample answer.*)

SOME SIMPLE ACTIVITIES FOR YOUR CLIENTS*

Arts and Crafts

Making a Collage   A collage is simply a group of pictures pasted together. Have clients tear or cut pictures from magazines and paste them onto construction

---

*Much appreciation is extended to the staff at Columbus State Institute (Columbus, Ohio) for these ideas.

paper. This activity can be modified to the different client needs. If a client cannot tear or cut, have the pictures ready or teach these skills. If a client has difficulty applying the paste to each picture, teach the client to spread the paste on the background sheet, and then just press the pictures on. The collage can also have a theme such as foods or animals.

Making a Mobile A mobile is a hanging object that usually has brightly colored pieces held together by string. People enjoy watching the objects move in a breeze. Suggested materials for making mobiles include:

Yarn balls

Seashells

Hardened dough

Scraps of material made into shapes such as animals, stars

Construction paper

Keys

Pieces of wood

Pine cones, acorns

Styrofoam, aluminum foil

Bells

The bases of the mobile can be made of such things as

Clorox bottles (cut)

Sticks tied together

Wicker paper-plate holders

Cardboard

Hard plastic lids

Heavy paper plates

Making Stained-Glass Windows The materials needed for making stained-glass windows are multi-colored tissue paper, white glue (diluted with water), plastic wrap, construction paper, and scissors. Have clients tear the tissue paper into small pieces and then paste them onto a sheet of plastic wrap. When the plastic wrap is covered with the colored tissue paper, frame it with the construction paper. Tape the stained glass window to the wall or to a window.

Working with Clay A sequence of activities that can be used to teach clients to work with clay (or Play Doh) is as follows:

Feel the clay.

Squeeze the clay.

Pound the clay flat.

Roll the clay into a ball.

Roll the clay into a snake.

Cut the clay snake with scissors.

Cut the clay snake with a knife.

Roll a ball of clay with a rolling pin and cut it with cookie cutters.

Make simple animals and objects out of the clay.

Gross Motor Activities

Gross motor activities involve the large muscle groups that are used in walking, throwing, running, and jumping. Many inexpensive items can be used in your activities with your clients. Games, such as Follow the Leader, can be adapted for use with this equipment. Also you can be creative, devising your own games and inventing your own equipment. Some examples of equipment and activities follow. Many of these games and activities were planned for the severely and profoundly handicapped in facilities with little recreational equipment. Ideally, of course, appropriate equipment and staff would be made available to clients.

Inexpensive Equipment Ideas for Gross Motor Activities The following is just a partial list of equipment that can be used to aid gross motor activities:

Auto tires

Trash and garbage cans

Balloons

inner tubes

Walking boards

Crepe paper

Golf club tubes

Rhythm sticks

Hoops

Yarn balls

Tennis balls

Masking tape

Carpet squares

Scooters

Ropes

Bean bags

Gallon milk containers

Large cardboard boxes

Cardboard and wooden shapes

Ladders

2×4s (for balance beam)

Bamboo poles

Barrels (drums)

Bowling pins

Obstacle Course An activity that helps people gain control over their bodies, obstacle courses can be fun and can be set up with different things that are easily available. For example:

Climb *on* a chair.

Jump *over* a block.

Go *around* a wastebasket.

Stand *in* a box.

Walk *through* the tires.

Walk *on* a 2×4.

At first the obstacle course might include only one or two items. As the clients learn what is expected of them, other obstacles may be added.

Walk a Straight Line Walking a straight line helps clients develop body control. It simply involves putting masking tape on the floor in a straight line (about ten feet long) and asking clients to walk on it. When a client can do this without any errors, the tape can be made into more complex designs (wavy line, square, triangle), and the client may also be asked to walk backwards on the tape.

Wheelchair Activities People in wheelchairs can also participate in gross motor activities. Some examples are the following:

a) *Wheelchair races.* Two clients of similar wheeling abilities race down a hall or sidewalk. These also may be done as relay races, with the clients passing an object to one another.

b) *Wheelchair basketball.* Group clients in a circle around a wastebasket and take turns trying to throw a ball into it. Keep individual scores and have clients count the number of baskets they have made. As the clients develop these skills, they can progress to using regular basketball equipment, and eventually into a game.

c) *Obstacle course.* Set up an obstacle course in a large open area, using readily available materials. This activity can be used to teach various concepts such as "go *through* the door, go *around* the chair, go *under* the broom laid across two chairs, go *over* the tape on the floor, go *between* two chairs."

d) *Wheelchair volleyball.* Put a string between two chairs to act as a net and use a balloon for a volleyball. As the clients' abilities improve, they may be introduced to a regular volleyball game.

Ball Activities   Throwing and catching a ball follows a progression of activities from simple to more complex. The following progression is an example:

a) Tossing and catching the ball oneself

b) Bouncing and catching the ball oneself, with
both hands
dominant hand
nondominant hand

c) Rolling the ball to a partner at a close distance, then gradually farther away

d) Throwing the ball underhand to a partner, gradually increasing the distance

e) Throwing overhand to a partner, gradually increasing the distance

f) Throwing at a target (trash can, basket)

g) Hitting the ball. (It is easier to hit a ball with the foot or hand than with other body parts or tools such as a bat, racket, or paddle.)

As clients develop throwing and catching skills, other games can be attempted. A version of basketball, for example, using a trash can for a basket, can be played by a small group of clients. Bean bags can also be used with clients instead of balls in these activities.

Activities to Help the Client Discriminate Sights, Sounds, and Smells

These activities are just a small sampling of things that can be done with your clients. Use your imagination to invent new activities of your own.

Shape Box   Shape boxes can be made from bleach bottles (or similar objects) or can be purchased. They are simply containers with holes of various shapes into which the same-shaped pieces fit. The object is to put the pieces in the correct holes, an activity that teaches the client to distinguish between different shapes, using visual and tactile (touch) senses.

Sound Boxes   Sound boxes are containers (empty milk cartons or soda pop cans) with different objects placed inside; so when someone shakes them, each makes a different noise. It is a good idea to have two sets of identical boxes, so the client can shake a box from one set and then try to match the sound by shaking the boxes from the other set.

Smell Boxes Smell boxes are similar to sound boxes, except that the containers (pill bottles make good smell boxes) contain things that give off odors rather than sounds. Some good smells to include might be chili powder, cloves, coffee, coconut, garlic, pepper, peppermint, and ammonia.

ACTIVITIES FOR THOUGHT AND DISCUSSION

A. Pick one or two of the activities just described to do with one of your clients. Report back to the class on

How quickly the client learned the activity

What changes you made in your approach to help the client enjoy the activity more

B. Think of three other activities that your clients would enjoy in each of these areas:

Arts and crafts

Gross motor skills

Discrimination of sights, sounds, and smells

Most of the activities discussed thus far are appropriate for severely and profoundly disabled individuals. Of course, recreation and leisure-time activities are also important for moderately and mildly impaired individuals. Although many of these people are capable of working and living in the community, they may run into some difficulty because of their poor recreational and social skills. They may also have difficulty finding or making use of the resources available to them in the community. These residents should be counseled on how to use their leisure time, and should be taught how to make use of community resources (bowling, swimming, restaurants, movies, miniature golf). At first this can occur in small, supervised groups, but as the clients develop more skills in this area, they should be encouraged to use these services on their own or with one or two friends.

These clients are also frequently interested in dating and should be instructed in the basics (calling the date, planning the event, budgeting money for the date, how to act on the date), if necessary. Clubs and dances for both men and women offer an opportunity to meet new people and practice social skills. Finally, hobbies are also a great interest to many people and should be encouraged.

SPECIAL OLYMPICS

One major event that occurs each year is available to individuals of all handicap levels. This is the Special Olympics, first started in the 1960s, in which children and adults compete in a variety of events. These include track and field (running, jumping, relay races), swimming and diving, bowling, broom hockey, volleyball, and many others.

Although the Special Olympics occurs once a year, many people spend months preparing for them. A major focus in many local recreation programs is getting clients in shape for the local, state, national, and international trials. The winners of each meet (local, state, national) have the opportunity to participate in the next level of competition.

The Special Olympics allows even those with several handicaps to compete with one another. Events designed for people in wheelchairs, for example, are an important part of the Special Olympics.

This yearly event is an exciting time for many people. The developmentally disabled individuals who participate are proud of their accomplishments. The Special Olympics provides the athletes with a chance to succeed and learn, to think of themselves not as disabled, but as competitiors. The Special Olympics allows people to feel good about themselves.

Sexuality and the Developmentally Disabled Person

10

Carol Winters Wish, Ph.D.
Joel Wish, Ph.D.

SEXUAL RIGHTS AND RESPONSIBILITIES OF THE DEVELOPMENTALLY DISABLED

Direct care staff are responsible for the developmentally disabled person's total well-being. Most of us feel comfortable in dealing with routine physical, educational, and social matters. When we discuss the sexuality of persons with developmental disabilities, however, we enter an area in which we often feel uncomfortable, and that sometimes we wish we could ignore. Nevertheless, it is an important topic and must be dealt with directly.

Society has only recently considered persons with developmental disabilities as deserving of their full human rights. This includes admitting that the developmentally disabled are also sexual beings. The issues of heterosexual relationships, homosexual relationships, birth control, sterilization, and masturbation must be considered. In short, we must maturely, humanely, and without prejudice examine the area of sexual rights and responsibilities of persons with developmental disabilities.

We have learned that the principles of normalization and deinstitutionalization attempt to provide the least restrictive, most normal lifestyles for our clients. Our goal is to prepare them to function as independently as possible. Often, however, we neglect to prepare the person with developmental disabilities to participate in normal social, affectional, and sexual relationships to the fullest degree possible. Indeed, many people might find the area of sex education for the developmentally disabled a frightening one. They may be unsure of the way such knowledge may change the developmentally disabled person's behavior. This area of instruction may also produce feelings of insecurity for us, as we come face to face with our own feelings toward sexuality. Yet when we do not offer basic social and sexual education, the developmentally disabled individual is open to increased social and sexual risks. These people are, in fact, often exposed to sexual information by television, magazines, and friends in schools, workshops, and residences. Our goal should be to help our clients function responsibly in social situations.

In summary, we must consider that two of the human rights of the person with developmental disabilities are the right to be recognized as a sexual being and the right to education concerning socially appropriate sexual conduct. The mentally retarded individual has the right to privacy, respect for his or her affectional and sexual relationships, and the right to accurate and understandable social (e.g., hygiene) and sexual education. We must respect such rights and provide socially appropriate and responsible means of sexual expression.

HUMAN SEXUAL DEVELOPMENT

Developmentally disabled persons experience the same patterns of sexual development as normal persons, but generally at slower rates. It is helpful, then, to remember that sexual development of people with developmental disabilities is much more *like* the development of others than it is different.

From infancy, humans begin to explore and discover their individuality, becoming increasingly aware of the pleasures and capacities of their own bodies and senses. Young children receive great pleasure from gradually mastering their bodies and becoming more capable in the skills of grasping, walking, running, toileting, and talking. The physical exploration of infants and toddlers is considered "sexual" to the extent that such exploration involves the search for pleasure. These early physical experiences assist the individual in developing a self-concept.

As young children become more social, they naturally become curious about their own physical experiences and those of peers and parents. If parents do not answer questions of preschool children honestly and directly, but respond with alarm, anger, or embarrassment, the children may learn to feel, without understanding why, that the human body is "dirty" and that pleasure derived from one's body is "bad." During the early school years, "dirty jokes" may relate to toileting and sex and often reflect misinformation and unsatisfied curiosity. Young children are usually curious about the bodies of other boys and girls, and it is not unusual for them to question the nature of conception, gestation, and birth. They may become frustrated and acquire negative feelings and attitudes toward sex, if their curiosities are met with silence, rejection, or abstract technical explanations that they cannot understand. Good sex education is concrete, honest, and direct from the start. It encourages and welcomes open expression of feelings and inquiry by children about their uncertainties.

In later childhood many children especially enjoy close friendships with other children of the same sex. They become interested in activities and people outside their families and begin developing the skills needed for adulthood. In both sexes, sexual behavior normally includes experiments with self-stimulation (masturbation) and sexual experimentation with same-sex friends. These behaviors are considered normal aspects in the development of sexuality.

After puberty, adolescents develop a growing ability for adult genital sexuality, as well as new aspects of their own personalities that accompany new sexual needs and drives and new social expectations and responsibilites. As normal adolescents develop sexual and affectional ties with peers, they also become more independent of the family.

Adult sexual development is finally attained once an individual has learned to experience a mutually satisfying sexual relationship with a willing partner. In our society this has usually been attained through marriage and parenting. Today, however, our society is showing more acceptance of those who choose to remain single and to adopt other lifestyles, such as living together without marriage, homosexuality, or single-parent families. Once again, we must understand each individual's sexuality in terms of his or her individual development. In considering appropriate sexual expression we must examine the individual's total needs and abilities and remember that all persons, regardless of developmental level, have the need for close relationships with others.

Sexual Expression and Developmental Level

In considering individuals we call developmentally disabled, we must understand an individual's sexuality in relation to his or her individual development. In considering appropriate sexual expression, we should examine the individual's total needs and abilities.

Many mildly retarded individuals achieve the full range of sexual expression as outlined above. These individuals generally prefer more concrete, traditional sex roles, although other roles may be learned. Generally speaking, mildly retarded adults are able to learn to successfully combine vocational and home-making roles. When providing sex education to mildly retarded people, keep in mind that they often benefit from lessons that stress the responsibilities they are to assume whenever they engage in sexual activity with another person.

Moderately retarded individuals are also generally capable of the full range of sexual activity and demonstrate an awareness of their sexual behaviors. If good training and guidance are provided, moderately retarded people can learn appropriate social, sexual, and affectional behaviors. Although adults with moderate retardation may hope for long-term loving relationships, these hopes typically lack depth of understanding. Their perceptions of sex-role differences are generally rigid because they lack an understanding of social and sexual behaviors and responsibilities. Generally, persons with moderate mental retardation need social-sexual education that covers the basics of a variety of skills, such as becoming involved in simple conversations, how to plan a date, and what to do on a date. In addition, these residents should also be introduced to the various birth control methods. Opportunities for role-playing and much repetition of the educational material should be made available. Thus, a man and a woman in a classroom setting may role-play a telephone conversation in which the man asks the woman for a date.

Persons with severe mental retardation may also understand some of the more basic sexual differences found in the mainstream of society. For example, their choices of play and work activities may demonstrate strong sex-role differences. The severely retarded may express their sexual desires through masturbation, as well as heterosexual or homosexual sex-play. The severely retarded person may possess poor control when sexually aroused, because of a lack of understanding. Severely retarded people are generally more desirous of immediate satisfaction and are less aware of the implications and consequences of their sexual behavior.

Profoundly retarded persons may be able to identify only the more superficial and the most obvious sex differences, such as hair length, dress, and pitch of voice. Sexual expression among these people is often directed toward themselves. Profoundly retarded individuals normally take pleasure in the sensations felt in their own bodies, usually through genital, oral, or anal stimulation. It must be stressed, however, that affectional relationships with peers and staff are very important for people with profound disabilities. These relationships may be simply expressed through caring, talking, touching, and holding, and may communicate concern and friendship.

General Considerations

The mentally retarded individual may be especially vulnerable to exploitation or attack. This may be due to poor thinking ability, inexperience, fear, or an attitude of general trust toward others. Therefore, we have a special duty to provide sex education and to teach the difference between appropriate sexual expression and exploitation. A first step in this process is to find out the beliefs, attitudes, and misconceptions the individual possesses prior to any sex education training.

The intellectual limitations of the retarded person require a concrete, clear, and personal approach to sex education. Special materials should include photographs, actual birth control devices, life-size or realistic models, and the like. The content of the course should be stated in basic and simple terms. Abstract concepts, such as *privacy* and *anatomy* ("plumbing"), must be translated into behavioral terms. Behavioral examples and role-playing are more important than the clients' being able to use terms correctly. There are tests, guides, and courses available for teaching sex education to persons with retardation (see "Social-Sexual Education Training Materials and Guides" in appendix B).

In determining what and how to teach about social-sexual matters, a major consideration is our attitude toward sex and the mentally retarded. If staff are overly protective toward their clients, they may not provide clear examples of what behaviors cause problems for the clients or others and what behaviors are permitted. Often, specific information may be withheld because of the mistaken idea that a retarded person will have no need of it. For example, if the mildly or moderately retarded individual is viewed as an "eternal child," he or she may not be taught about adult sexuality. If human feelings are not to be expected of the retarded, relationships between retarded adults may be ridiculed, referred to as "cute," or perhaps forbidden. With mildly and moderately retarded persons, adolescent types of relationships are often prolonged as a result of these attitudes. In short, well-meaning caretakers sometimes fail to see that the question is not whether sex education *should* be provided (for all humans will experience sexuality), but *how* these individuals will learn about sex. The issue is whether they will learn and respond to misinformation and attitudes learned from unreliable sources, or whether they will be offered concrete, accurate, and complete information to meet their developmental needs.

MARRIAGE, INTIMACY, AND PARENTING

Many mildly and moderately retarded persons express a desire for marriage and a family. The role of the helping person is to respect the dignity of these desires and to promote a thoughtful and realistic assessment of the probability of successfully achieving these goals. Discussions of financial responsibilities, budgeting, and vocational development are crucial for couples who want to marry. Opportunities for open discussion with supportive counselors should be available before and after marriage as needed, sometimes on a regular basis.

In addition to the explanation of responsibilities in marriage, the topics of intercourse, pregnancy, and childbirth must be carefully and concretely presented. It is crucial to discourage the romantic myths concerning love, marriage, and parenting, and to concentrate instead on the physical and emotional needs of infants and their parents. Role-playing is an excellent technique for teaching about marriage and parenting. The financial and emotional responsibilities of marriage and parenting also need to be discussed, as couples are helped to assess realistically their abilities to be successful parents.

In our culture we have typically been taught to pair marriage with the expectation that a couple will bear and raise children. As a result, some people oppose marriage between the developmentally disabled (even though they are capable of loving and fulfilling relationships), simply because they feel the retarded couple is unable to care for a child responsibly. Marriage, however, does not always involve childbirth. Training for retarded persons who desire to be married but are not capable of responsible parenting must stress the differences between marriage and parenting. The developmentally disabled should also be exposed to the knowledge that alternative lifestyles can be as rewarding and fulfilling as traditional marriage relationships.

While there is not a strong relationship between intelligence and success-ful marriage, childless marriages of retarded persons are more apt to succeed. A goal of social-sexual education is to help the individual realistically evaluate his or her ability to meet demands that arise when one exercises one's human rights to marry or to perform other social-sexual functions. Once we acknowledge the right of the retarded person to be sexually active, we must also acknowledge his or her right and responsibility to use birth control. The pros and cons of the available birth control methods can be presented concretely, using real examples. The sexually active retarded person also has the right and responsibility to know about venereal disease. Training should include the outstanding symptoms of VD; where, why, and when to receive treatment; and preventive measures.

One of the most controversial topics in the field of mental retardation is the issue of sterilization. Problems with third-party consent (that is, a parent or guardian's signing consent forms for individuals considered unable to care for themselves) have resulted in lawsuits and legal battles. The question of sterilization has complex legal and moral implications relating to individual freedom and the right to the least restrictive treatment. Included in such a decision are the family's rights and responsibilities, as well as those of the individual and society at large. Decisions regarding sterilization are best made on a case-by-case basis.

It is very important for staff to be respectful and objective when dealing with sexual issues. When a staff person feels that he or she cannot discuss a topic nonjudgmentally, the client should be referred to another person for training. Social-sexual training will not always be easy or without moments of discomfort. Nevertheless, objectivity, openness, honesty, and the ability to recognize and admit one's own limitations are crucial.

MASTURBATION AND HOMOSEXUALITY

Masturbation and homosexuality are often behaviors of concern, especially to staff in institutions and other residential settings. The most useful response to these concerns is to ask, "What alternative sexual behaviors are available?" In many settings, masturbation and homosexuality are the only sexual outlets available. Self-stimulation (e.g., masturbation, rocking) is very common where surroundings are boring and there is little else to do. Providing other sources of stimulation can prove helpful. For example, positive interaction with staff may be very effective in curbing excessive masturbation. Where staffing is too low to permit extensive one-to-one staff-resident interaction, improved lighting, colorful surroundings, group activities, mobiles, films, new textures, and other equipment and activities will provide external stimulation and may reduce the need for self-stimulation. Both masturbation and homosexuality are legitimate means of sexual expression. Training in both areas should focus on the issue of privacy and, in the case of homosexuality, upon consent between willing partners.

A setting in which only men or only women live, with few sources of outside interest and few sexual outlets, may encourage homosexuality. *Situational homosexuality* is common in all forced sex-segregated settings. We can respect the sexual rights of residents by making available alternative sources of sexual expression, not by restricting sexual expression altogether. Providing a more stimulating environment and frequent opportunities for opposite-sex interaction will decrease the incidence of homosexuality. Dances, coed recreation, coed leisure time, coed work, and therapy all decrease the necessity for homosexuality and masturbation due to lack of other programming or opportunities to form opposite-sex friendships. It is very important to note that aside from the amount and type of programming available, there are developmentally disabled persons who prefer homosexual relationships, and that homosexuality is a legitimate form of sexual expression.

PRIVACY

A most important concept to teach retarded persons is that of privacy. As we teach retarded persons that sexual feelings and behaviors are O.K., we must let them know where and when such behaviors are acceptable. Restrictive settings often limit privacy unnecessarily by providing doorless bathrooms, unlockable doors, group dormitories, sex segregation, and regimented activities. In addition, continuous supervision by staff deprives the individual of privacy. Thus, as with a goldfish in a bowl, we note "unacceptable" behaviors that might not be observed in settings where more privacy is available. In addition, where we do not provide for individual privacy, we prevent the development of a sense of self-pride and independence. It is just as crucial that retarded persons living in the community have a well-developed concept of privacy, since sexual misconduct can, and frequently does, lead to their return to the institution or removal to a more restrictive environment.

SOCIAL EDUCATION

Finally, emphasis should be placed upon teaching skills pertaining to hygiene (e.g., caring for menstrual periods, cleaning genital areas, use of deodorant, body functions); alcohol and drug use (or abuse); saying "no" to unwelcome sexual advances; and the development of relationships (including meeting people and how to spend leisure time alone and with others). Again, role-playing of appropriate greeting and dating behavior is an effective teaching technique.

SUMMARY

The central theme of this chapter is the recognition that sexuality is a normal part of all human functioning, even among those who have developmental disabilities. We have outlined the general sequence of sexual development and noted how people with various levels of developmental disability may or may not differ from the sequence. We have discussed the types of social and sexual education necessary to prepare a retarded individual for the least restrictive expression of his or her sexuality. Masturbation and homosexual behavior are recognized as normal behaviors that may increase in environments where there are few sources of stimulation or few alternative sexual outlets. Increasing the interest and stimulation of surroundings and increasing opportunities for normal heterosexual contacts tend to decrease masturbation and homosexual behavior. Increased privacy provides opportunities for clients to learn other normalized behaviors.

We have noted that some seemingly "sexual" behavior is actually a normal response to dull, boring, sex-segregated environments. Staff are in a unique position to increase the stimulation of the environment by providing a change and increase in activities. Providing coed activities and social occasions (e.g., mealtimes, recreation and leisure-time activities, therapy sessions, work situations) are good adaptive measures. We must also appreciate, however, that masturbation and homosexual behavior are appropriate outlets for sexual expression.

Sex education must be ongoing and appropriately attuned to the individual's developmental needs. Concreteness, structure, and repetition are often necessary for teaching developmentally disabled persons. The helping person should foster an honest expression of feelings and attitudes among those he or she teaches by honestly and objectively presenting information. It is important for the educator to respect the dignity and to consider the normal needs of the person being taught. Any suggestion of "wrongness," "badness," or "evil" should be avoided. An attitude of ridicule or condescension will also thwart or damage the retarded person's confidence and healthy sexual development. The sex-educator should know the students—that is, their knowledge, misinformation, degree of understanding, and attitudes prior to and at the close of formal training. The helping person has a responsibility to teach not only social-sexual behaviors, but also the consequences and implications of these behaviors. Like everyone else, when developmentally disabled people assume their social and sexual human rights, they must also learn to exercise personal and social responsibility for their actions.

QUESTIONS

10-1. What can you do to make your clients' living situations more socially and sexually humane and normal?

10-2. You overhear another staff person ridiculing a male and female client who are holding hands while watching television. Should you say anything? If so, what?

10-3. A twenty-five-year-old female client confides that she is in love and wants to marry a man she met at her sheltered employment. Should you respond? What issues might you consider? How should these issues be presented?

10-4. A twenty-year-old female resident confides that she has had sexual relations with several men. She appears to have little understanding of reproduction or knowledge of birth control. What do you do?

10-5. A sixteen-year-old female resident complains that she is being sexually harassed on her way to work. How do you handle this situation?

10-6. Aides complain that a middle-aged male resident of a state institution masturbates "constantly." You are asked what to do. How do you respond?

10-7. Two fifteen-year-old male residents of a state institution are observed in mutual sexual stimulation in the dayroom of their ward. Some staff ask permission to punish the boys by putting them in isolation rooms, claiming that these boys are homosexual and should be punished. How do you respond? Is the boys' behavior normal or abnormal? How does the concept of *privacy* relate to this incident?

10-8. The mother of a thirteen-year-old girl with Down's syndrome tells you she thinks her daughter should be sterilized immediately. What moral and legal issues do you suggest the mother to consider?

10-9. A twenty-year-old client in a sheltered workshop confides that she wants nothing more than to have a baby. What is your response? What issues do you encourage her to consider? How do you present these issues?

10-10. A sixty-year-old female resident of a state institution repeatedly takes off her clothes and walks around naked. She has lived in the institution since she was twenty. You are asked to explain and change her behaviors. What do you do?

10-11. A twenty-year-old male resident of a community living arrangement is eager to date a woman he met at church. He tells you he wants to marry her and have children. How do you respond? What issues do you encourage him to talk about? How do you support his desire for a more normal social and living situation?

The Developmentally Disabled Child and the Family

11

A developmentally disabled individual's handicapping conditions generally continue throughout life; therefore, his or her influence on parents, brothers, and sisters will also continue throughout life. This is especially true of people who are more severely disabled. Someone who is severely mentally retarded and cerebral palsied might cause more of a disruption in the family's routine than someone who has epilepsy.

Each family reacts to the birth of a handicapped child in a different manner. All families must certainly be disappointed when their child is born with a disability. However, how the family copes with disappointment may vary. This disappointment may quickly turn into sorrow or grief. In some families, this sorrow and grief may last for many years, while in others such accidents of life are accepted more quickly. Some families are able to continue as a unit, while in others the birth of a handicapped child creates so much stress that marital and family problems develop and perhaps even lead to divorce.

EFFECTS OF DEVELOPMENTALLY DISABLED CHILD ON THE FAMILY

The introduction of a developmentally disabled child into a family has the potential of changing the family's lifestyle. These changes begin to occur whenever parents learn that they have a handicapped youngster, shortly after the birth of the child if the disability is immediately obvious to the physician (as in the case of Down's syndrome), or perhaps when the child is several years old. Many children's disabilities, for example, are diagnosed when they enter school.

Still other children may develop normally until an accident or illness causes brain damage that results in a developmental disability. In each case family expectations and behaviors change.

A Change of Expectations

Whenever parents plan to have a baby, they generally have expectations and goals for that child. Some parents think about their unborn child's future career (doctor, mechanic, police person, nurse) or expect great artistic or athletic accomplishments from their child. When a child is born with a handicap, many of these dreams are shattered. Parents who go through such an experience often feel helpless and believe that their child's future is hopeless. Counseling is often helpful in assisting parents to recognize their child's potential and to accept all that the child is capable of giving (love, affection, devotion, and the normal pleasures of childrearing).

After becoming more familiar with the handicapping conditions, many parents are able to change what they expect their child to accomplish in life and to support their child in reaching more realistic goals. Groups of parents with disabled children can often help each other through the different life crises and changed expectations that arise. These support groups also allow parents to express their frustrations and concerns to people who "really understand" because they have lived through similar experiences.

Changes in the Family's Lifestyle

A developmentally disabled child can often create burdens on the family that a normally developing child does not cause. These burdens can affect all members of the family and may require a rethinking of the family's goals. Some of the effects that having a developmentally disabled child may have on a family are as follows:

Time Parents all recognize that if they want children, they will have to devote time, energy, and money to raising them. How is this different for a developmentally disabled child? A developmentally disabled child often requires the parents to spend more time and devote more energy in their physical care and training than does a normal child. For Example, a normal child learns to eat with a spoon or fork, drink from a cup, and use the toilet, all before the age of three. A severely or profoundly mentally retarded child might not learn these skills until years later. In addition, the parent may have to attend some special training courses to help them teach their child these skills more effectively.

From this example it can be seen that parents of a disabled child often work longer and harder in this teaching even the basic skills to their children. And

---

Later in this chapter we will discuss how the family reacts to a developmentally disabled child.

developmentally disabled children often take longer than their normal peers to achieve many other life skills as well, increasing their dependency on their families. This dependency means that parents often must devote more time to the care of their developmentally disabled child than is necessary for other children.

Sometimes, the burdens of raising a developmentally disabled child are so great that parents turn over many responsibilities to brothers or sisters. This often limits the latter's social activities and at times may lead to resentment toward their "special" brother or sister. Thus, the presence of a developmentally disabled child can change each family member's activities, as well as the general atmosphere in the home. There are, however, some possible positive effects of having a developmentally disabled child in the family. The family might develop a common purpose through helping the disabled child develop to his or her fullest potential. This might open up channels of communication between family members, as well as increase their feelings of self-worth.

Money Raising a developmentally disabled child is often more expensive than raising a nonhandicapped child. Some of these children have medical problems that require frequent attention, and others need special diets, which may be both expensive and time-consuming to prepare. Other children may require special equipment (wheelchairs, leg braces), medicines, or professional services (physical therapy, speech therapy, psychotherapy) that may burden the family financially. Thus, especially in families with limited incomes, a developmentally disabled child's needs may largely determine the way the family budgets its money.

Social Life Having a developmentally disabled child in the family may limit family members' social lives to some degree. Often neighbors and friends do not understand what a developmental disability is and may withdraw from the family. In other cases the family itself may feel uncomfortable or embarrassed by their child and may avoid contacts with friends and neighbors. Still another reason for a possible decrease in interaction with friends is that the time and money involved in raising a handicapped youngster may limit social opportunities.

Brothers and sisters of a developmentally disabled child may also find their social outlets limited because they must help in raising the child. Thus, a developmentally disabled child may indirectly limit the family's social life by absorbing so much of its time and energy.

Of course, not all families are affected in these ways. Many continue their social life and include their developmentally disabled child in these activities. This is a much healthier approach for the family as a whole.

Parents in Later Life All families change over time. As normal children grow older, they spend more and more time away from home and need less of their parents' attention and guidance. Eventually they move out on their own. Parents normally expect this to happen and adjust their lifestyles accordingly. In families with a severely disabled person at home, however, parental roles may not change over time (Farber 1959). Parents may feel responsible for caring for their son or daughter through his or her adult years. Also they may feel guilty about placing their son or daughter in an appropriate residential setting (or even about considering it), regardless of the fine quality of care available in many communities. If their child is placed in a residential setting, some parents may be overly concerned about their son or daughter's welfare, feeling that no one can care for their disabled offspring as well as they can. Often, listening to parents, reassuring them, and keeping them up to date (with the client's permission) on the client's programming help them overcome this problem.

FAMILY REACTIONS TO HAVING A DEVELOPMENTALLY DISABLED CHILD

Family Support Systems

Although there are some general patterns that families go through when they discover they have a developmentally disabled child, different families handle the news in different ways. For example, most families experience shock when informed that their child has a handicap, but the length of time that this reaction persists, as well as its severity, varies from family to family. How each handles the news depends on several different support systems.

Family and Friends Parents generally have an easier time accepting their child's handicaps if their family and friends continue their close relationships and also accept the developmentally disabled child into their circle. Often, however, one's neighbors and relatives do not know how to react when seeing the parents of a mentally retarded child and may avoid them. When this occurs the parents may feel rejected and may blame the child. Thus, it can be seen how important it is for new parents to have the support of relatives and friends.

Religion Many people in times of stress gather strength from their church or synagogue. Religious organizations often provide support for the family and usually accept the handicapped youngster into the congregation. Different religious beliefs also provide the family with some new thoughts and personal strengths that help it cope with the stress of having a developmentally disabled child (Farber 1959).

Parents' Organizations and Support Groups Many parents can receive support from knowing that other families are experiencing similar types of stresses. In addition, parents can learn about new coping techniques from each other, as well as possible crises that may occur as their developmentally disabled children grow older. Parent groups have also been the major driving force in the development of new community programs for the developmentally disabled.

Counseling   Counseling often provides an important support function to many families of developmentally disabled individuals. The form the disability takes is often not as important as the way parents, brothers, and sisters view it. Counseling often helps family members change their attitude from an initial feeling of overwhelming catastrophe to a realistic acknowledgment of having to make changes in lifestyle, while still being able to enjoy life, themselves, and others.

Some Typical Family Reactions

There are a variety of ways that parents and siblings (brothers and sisters) react to having a developmentally disabled child. For some people these reactions last only a short time, and then the family moves into a stage of accepting the child's handicaps and then toward adjusting their expectations and plans. Some families, however, do not reach a stage of acceptance. Members of these families often deny that the child has a handicap, may reject the child, or may continue to search for someone to tell them that their child is not handicapped. In some cases the family may even fall apart because it cannot deal with the stresses of having a developmentally disabled child (Wolfensberger and Menolascino 1970). Let us now look at some reactions that parents may have.

Denial   The first reactions that many parents have when informed that they have a developmentally disabled youngster is to deny that the problem exists. These parents are often convinced that the psychologist, physician, or other professional making the diagnosis is wrong, or that the handicap is not as severe as the professional believes. Some parents may spend a fortune trying to find a magical cure for their child.

It is not our job to dispel parents' hopes for their child's future. Hope is a strong force, which motivates parents to strive to help their children as much as possible. Unfortunately, parents may also use hope as a way of blinding themselves to the realities of their child's condition. When this occurs, parents may also disregard the advice and programs available to them and their child.

Guilt   Once parents acknowledge that their child has a handicap, they may feel guilty, thinking that it was something they did that caused the disability. (In most cases it is difficult to determine what caused the disability.) Some parents may feel that "God is getting even" with them for something they did years before.

Guilt may occur at different times in the life cycle. Parents may feel guilty when they decide to have their adult son or daughter move into a group home. These parents may feel that since they brought a handicapped child into the world, they are responsible for taking care of that person for the rest of their lives. As a result, they may feel guilty over giving some of this responsibility to others. Parents often overcome these feelings when they receive accurate information about the handicap, or when they discuss their concerns with other parents or through counseling.

Depression Feelings of depression often occur along with guilt. Some parents may go through a grief process, as if having a developmentally disabled child is the same as having a child die. Parents and siblings who become depressed over having a developmentally disabled child often ask, "Why me, what did I do to deserve this?" Feeling very sorry for the handicapped child and the fact that he or she cannot achieve all that was hoped for, they may overly blame themselves if they make a mistake in caring for or teaching the child.

Parents may also become depressed occasionally because having a developmentally disabled child may mean changing some of their plans. Others may become depressed when thinking of the extended portion of their lives that will be devoted to parenting beyond what would be necessary for a normal child. Depression may also occur when the parents compare their original expectations for their child with more realistic expectations. They may be saddened when they think about what they had hoped or expected from their child (before the child was born or before the disability occurred), and what the child is really capable of accomplishing.

If parents' depression is temporary in nature and results from a variety of life stresses, it is generally a normal response. If the depression is of long duration, however, or results in a parent's inability to cope on a day-to-day basis, then professional help should be sought.

Embarrassment Parents often view their children as extensions of themselves. When the child behaves inappropriately or "looks funny," they may become embarrassed and may feel awkward in public. Brothers and sisters may also experience embarrassment for the same reasons and may feel uncomfortable in bringing their friends home to visit. When parents feel embarrassed because they have a developmentally disabled youngster, they may do some unusual things to hide their child's handicap. For example, a mother whose child goes to a preschool drove the child there daily instead of allowing her child to go on the school bus. Why? Because the school bus has a sign that reads, "County Program for the Mentally Retarded," and the mother did not want the neighbors to know that her child might need some special help.

Being provided with appropriate knowledge about handicaps, as well as involvement in support groups or counseling may prove helpful in overcoming these feelings of embarrassment. For brothers and sisters of developmentally disabled children, *sibling groups* allow children to share their feelings about their handicapped brother or sister and may provide new ways of coping.

Jealousy and Resentment   Family members, especially brothers and sisters, may be jealous of the developmentally disabled child in the family—not jealous of his or her handicap, but jealous of all the attention the child receives from the parents and of special programs the child attends. In addition, a developmentally disabled child may require expensive care and treatment, limiting the amount of money available for the nonhandicapped children's social life. Thus, in addition to being jealous, brothers and sisters may also resent the developmentally disabled child for imposing limits on their own activities.

Adjustment and Acceptance   Families undergo the emotions we have discussed (as well as other reactions) to varying degrees over the years they care for their developmentally disabled child. These emotions are often normal reactions to stressful situations. They become abnormal only when they predominate in the interactions between family members.

Most families accept the fact that they have a developmentally disabled youngster and adjust their lives so that the youngster's needs, as well as other family members' needs, can be met as much as possible. Even members of these families, however, may experience guilt, depression, anger, embarrassment, jealousy, or resentment from time to time. These emotions surface especially during stressful times for the family. Thus, a father may resent having a developmentally disabled child when he has to turn down a promotion that would require the family to move to a city with poor services for the developmentally disabled child. Usually, these feelings last a short time, and the family members can resolve the conflict on their own. When the resentment or other emotional reaction, however, affects the lives of family members for a long period of time, professional help should be sought. Psychologists, pastoral counselors (clergy), social workers, and psychiatrists are available to assist families with these problems.

QUESTIONS

11-1. In what ways does a developmentally disabled child affect his or her family?

11-2. What are some possible reactions that parents may go through when they find out they have a developmentally disabled child? Do all parents resolve their conflicts?

11-3. What can families do to get assistance in coping with the additional stresses that may occur when they have a developmentally disabled child? Are these services available in your community?

Staff Burnout: What It Is and How to Avoid It

12

Working with developmentally disabled individuals can often be emotionally draining. Staff often devote much time and effort in providing their clients with quality care and good programming. Progress, however, is often slow for our clients, and we often expect more from them and from ourselves. This proves to be frustrating to both clients and staff. In addition, superiors often apply pressure on staff to produce quick solutions to clients' problems, yet often provide little recognition for what one does accomplish. This can add to the frustrations of the job. These and other factors often lead to *staff burnout* (Lamb 1979).

Burnout occurs when staff have difficulty coping with the emotional stress of their jobs. It is characterized by a loss of concern and feelings for the people with whom they work. Burnout may even cause staff to treat clients in a detached or dehumanized manner (Maslach 1976).

SIGNS OF STAFF BURNOUT

When staff experience burnout, a variety of behaviors occur which do not usually happen all at once, but develop gradually. If you can recognize these behaviors as they are developing, then you may be able to do something to prevent them from becoming full-blown. (Several recommendations will be included at the end of this chapter to help you prevent or minimize the effects of burnout.) Through a variety of symptoms we can recognize burnout in direct care staff (Munro 1980).

Less Commitment to the Job   An early sign of burnout is when staff start to refer to their clients by using offensive names like "crazies," "retards," or "low functioning." This is a common sign that staff are beginning to "turn off" to their clients. When this occurs staff often prefer controlling their clients through medical means (medication to control their clients' behavioral problems), rather than using programming or other techniques that would involve spending time and interacting with the clients. When they lose their committment to the job, they also begin to dread coming in to work and may even perceive the clients as purposely making their job more difficult.

Less Attention Paid to Clients   When staff begin to experience burnout, it is not unusual for them to spend as much time as they can in the office or in the break room. They may find excuses to leave the clients to get various items, or they may even call in sick to avoid work. When doing the routine chores with clients, staff who are experiencing burnout may avoid touching or looking at them and may attempt to socialize with other staff rather than work directly with clients.

Signs of Depression   Staff who are experiencing burnout may look and act depressed, or may tend to cry or yell very easily. The least frustration may lead to client abuse, either verbal or physical. These staff may also be very resistant to suggestions from others, but at the same time they may accuse coworkers and administration of not caring about them or the clients.

Difficulty Working with Others   As burnout progresses, staff become uncooperative in developing or carrying out client programs. Their negativism begins to focus on their coworkers as well as the clients. They may talk behind coworkers' backs and generally find fault with everything around them.

Decline in Physical Health   Staff who experience an advanced degree of burnout may develop ulcers, migraine headaches, or other serious illnesses and may have a great deal of difficulty sleeping. To get through the day, as well as cope with their jobs and other difficulties, these staff may turn to tranquilizers, alcohol, or other drugs.

Increased Marital or Other Personal Problems   The stress that the staff feel at work often extends to their home life. This may lead to fights and arguments with their spouses and children and may eventually lead to divorce.

Clients' Reactions   As staff spend less time with their clients, the latter may react in a number of ways to get the staff's attention. For example, clients may become more abusive and aggressive. Usually, settings where staff are burned out also have clients who are in need of much care, attention, and programming.

HOW DOES BURNOUT DEVELOP?

Many factors within the work setting can contribute to burnout. Some may have to do with how the workplace is set up administratively, while others may result from working intensely with clients. Following are some common factors that may lead to burnout (Munro 1980).

Not Enough Consultation with Professionals Professional staff rarely work an eight-hour shift directly with clients. They often come and go at their convenience, rather than being available when the clients or the direct care staff need their input. If this is happening in your facility, discuss the problem with your supervisor and the professional staff person so that he or she can be more available.

Poor Communication Poor communication may occur between personnel on different shifts, between staff working on the same shift, and between direct care staff and the administration and professional staff. When this occurs, staff often feel at a loss in knowing what to do with their clients. They may also feel that they have no important decision-making powers that involve client programming.

Poorly Defined Staff Roles If staff are unaware of their specific job responsibilities (a problem often due to poor communication), they may resent being requested to perform other duties that they had considered outside of their realm.

Working Double Shifts Constantly Because of high turnover and call-offs in the human service field, direct care staff are often required to work double shifts. When this occurs frequently for any individual, there is an increased likelihood that that person will burn out quickly (Pines and Maslach 1978). When people frequently work sixteen hours straight, they often begin to see their job as an unbearable chore, having lost much of the free time they had previously used for enjoyable pursuits.

Working with Clients Whose Progress Is Often Slow and Difficult When working with severely handicapped clients, staff must often invest much energy, concern, and patience. Naturally, staff members as well as the administration hope that such clients progress quickly and without problems. When their clients' progress is slow, staff often become discouraged and may view this as a personal failure. When these frustrations occur frequently, especially if the administration is telling the direct care staff they are unsuccessful, there is an increased likelihood of burnout. Staff working with severely handicapped individuals should be aware that progress is possible, but should also maintain realistic expectations of their clients' prospects (Lamb 1979).

COPING WITH ON-THE-JOB STRESS
TO PREVENT BURNOUT

Many things can be done, on and off the job, to assist you in coping with job stresses. If you can more effectively cope with these various stresses, then you may avoid becoming burned out.

On-the-Job Changes

A Team Approach   When all staff are working together, including professional staff, there is a feeling of mutual support and common purpose. This may require more frequent interaction between direct care staff and professional consultants and more frequent communication of experiences with clients. This allows other staff to act as a sounding board for the job frustrations of the direct care staff.

Sharing the Client Load   It is generally recognized that working with the more severely handicapped clients may be more emotionally draining for staff than working with other clients. This may lead to increased burnout among these staff members. Sharing the load of working with these more handicapped individuals (by rotating the staff between wards), may remove some of the pressure by making staff's jobs more varied and stimulating (Pines and Maslach 1978).

Time-Outs   Staff should be allowed to withdraw temporarily from direct patient contact from time to time. This should be worked out on a scheduled basis by the supervisor to insure appropriate coverage of the clients. Time-outs would allow staff to change their environment for short periods of time during the week, perhaps by rotating cleaning, office, or medication responsibilities. In this way individuals' work routines are broken to some degree, and the quality of their time spent with clients may be improved (Pines and Maslach 1978).

Staff Bull Sessions   It has been found that in addition to regular case conferences, front-line personnel benefit from regularly scheduled bull sessions (Munro 1980). The agenda of these meetings should include a period of time for informal socializing where staff can develop solutions to common problems and provide support for their coworkers. It has also been found that when staff are involved in a retreat, leaving the work situation for an afternoon or a day to discuss work issues, their degree of burnout decreases.

Things You Can Do

1. Be aware of work stresses and recognize the signs of burnout.

2. Set realistic goals for yourself and for your clients. Do not try to accomplish the impossible. If you have clients with whom you are having difficulty working, discuss the problems with your supervisor.

3. Talk your frustrations out with your supervisor, especially when you are beginning to feel overwhelmed. Do not try to make these "bitch" sessions, but rather provide a realistic appraisal of your abilities and difficulties so that solutions to these problems may be developed.

4. Attend staff training programs. No one knows everything there is to do with the client. It is often helpful to attend staff training programs where new techniques are introduced or where old techniques are reviewed.

5. When your shift is over, leave your work behind. Fill your leisure time with things you enjoy doing. Your life involves more than just work, and one way to fight off burnout is to enjoy the rest of your day as much as you can.

6. Become involved in some physical activity after work. It has been found that daily jogging or other physical activity is a good way to separate your personal life from your work life and thus a good way to prevent burnout.

QUESTIONS

12-1. What is burnout? Have you ever experienced it?

12-2. What can you do to prevent yourself from burning out?

12-3. What constructive suggestions do you have for your place of work to help staff prevent burnout?

GLOSSARY

Abstract thought — Being able to generalize from experience.

Adaptive behavior — Ability to cope with social expectations.

Affect — Feelings, emotions.

Akinetic — Type of seizure where the person falls forward or backward.

Ataxia — Usually associated with cerebral palsy; inability to maintain normal balance, posture, and stability.

Athetosis — A form of cerebral palsy characterized by slower, wavelike movements of the hands and feet and often in the muscles of the face and tongue.

Aura — A subjective experience such as taste, sound, sight, or smell that precedes and marks the onset of an epileptic seizure.

Autism — A disorder of communication and behavior in which a person has limited ability to understand, learn, and participate in social relationships.

Backward training — Also called *backward chaining;* training a new behavior in the reverse order of the way it is usually performed (see *dressing programs*).

Behavior modification — Using the principles of behavior technology to systematically and intentionally modify behavior.

Cerebral palsy — A developmental disability due to central nervous system damage where a person has difficulty controlling his or her muscles.

Chromosome — Small particles found in pairs in all the cells of the body, carrying the genetic characteristics transmitted from parent to child.

Chronic — Marked by a long duration.

Clonic phase — Part of a grand mal seizure where jerking movements are present.

Communication — A process by which information is exchanged between people.

Conception — The act of becoming pregnant.

Concrete thought — When a person's thoughts deal with the immediate experience rather than generalizing.

Constrict — Draw the muscles together.

Continuous reinforcement — Used in training new behaviors; reinforcing each emission of the desired behavior.

Dehumanization — The process of treating people in less than human ways.

Deinstitutionalization — The process of moving people from institution to community.

Denial — Refusal to admit the truth.

Developmental disability — Term used to refer to specific handicapping conditions, including mental retardation, cerebral palsy, epilepsy, and autism.

Developmental model — A thesis that proposes that mentally retarded people go through the same sequence of developmental stages as normal individuals, but at a slower rate.

Developmental period — The first twenty-two years of life.

Discriminate — To distinguish one object or behavior from another.

Environment — Surroundings.

Epilepsy — A developmental disability in which an individual has seizures.

Extinction — A behavior modification technique in which a client does not receive reinforcement when the target behavior occurs, but is ignored for inappropriate acts.

Fetus — Unborn child in the mother's uterus from eight weeks after conception to birth.

Fine motor — Behaviors that involve small-muscle groups, reaching and grasping, for example.

Genitals — Sexual organs.

Gestation period — Period of pregnancy.

Grand mal seizures — A form of epilepsy characterized by a sudden loss of consciousness and intense muscle spasms.

Gross motor — Involving the large muscles of the body, walking, running, or throwing, for example.

Group home — A community residence where a small number of people live, involving varying amounts of staff supervision.

Habilitation — The process of training and assisting people to become as independent as possible.

Heterosexual — Interested in relationships with members of the opposite sex.

Homosexual — Engaging in sexual activities with a member of the same sex.

Impairment — A condition that limits one's abilities in certain areas.

Individualized Habilitation Plan (IHP) — A program plan developed for each client, which must be reviewed and updated periodically, noting client progress.

Institution — As used in this book: a state-supported facility of over 500 residents where people are placed to live for extended periods of time.

Interdisciplinary team — All those people who are working with a client and planning and carrying out programs together, a process that involves open communication between staff.

Intermittent reinforcement — Used in maintaining behaviors; the desired behavior is reinforced on a prearranged schedule, but not every time it occurs.

Masturbation — Self-stimulation of the genitals to achieve orgasm.

Mental retardation — Significantly subaverage general intellectual functioning, occuring concurrently with deficits in adaptive behavior and originating during the developmental period.

Modeling — A behavior modification technique by which a staff member shows a client how to perform the desired behavior.

Normalization — Making experiences available that are as close as possible to the rest of society, with the expectation that if people are provided with normal experiences, they will display more normal behavior.

Peer — Having an equal standing with someone else.

Perceptions — Observations, one's point of view.

Perinatal — During birth.

Petit mal seizures — A form of epilepsy marked by brief but frequent attacks of impaired consciousness, staring, blinking of the eyes, or loss of normal posture.

Physical guidance — A behavior modification technique by which a client is assisted through the desired behavior.

Postnatal — After birth.

Prenatal — Before birth.

Psychomotor seizures — A form of epilepsy that results in disordered, bizarre, or hallucinatory behaviors.

Reflex — An act performed involuntarily as a result of a nervous impulse.

Reinforcement — Any condition that, when presented after the occurrence of a specific behavior, increases the likelihood of that behavior's recurring.

Resentment — A feeling of displeasure about something.

Self-care activities — Bathing, eating, dressing, toileting.

Shaping — A behavior modification technique by which a client is reinforced for progressively closer approximation of the goals.

Siblings — Brothers or sisters.

Sign language — A system of hand gestures used for communication by people with hearing impairments.

Social reinforcers — Praise, smiles, and pats on the back used to increase the frequency of a behavior.

Sterilization — An operation to remove an individual's capacity to produce offspring.

Task analysis — Breaking down a task into its component steps so that it can be more easily taught.

Time-out — A behavior modification technique used to reduce the frequency of a certain behavior.

Tonic phase — Part of a grand mal seizure in which the muscles become rigid.

BIBLIOGRAPHY

Adams, M. *Mental Retardation and Its Social Dimensions.* New York: Columbia University Press, 1971.

Baroff, George S. *Mental Retardation: Nature, Cause, and Management.* Washington, D.C.: Hemisphere Publishing Company, 1974.

Bensberg, G. J. "Helping the Retarded in Language Development." In *Teaching the Mentally Retarded.* Atlanta: Southern Regional Education Board, 1965.

Blatt, B., and Kaplan, F. *Christmas in Purgatory.* Boston: Allyn & Bacon, 1966.

Conroy, J. W. "Trends in the Deinstitutionalization of the Mentally Retarded." *Mental Retardation* 15 (1977), no. 4: 44-46.

"Declaration of General and Special Rights of the Mentally Retarded. *Mental Retardation* 7 (1969), no. 4.

Dempsey, John J. *Community Services for Retarded Children.* Baltimore: University Park Press, 1975.

deSilva, R. M., and Faflak, P. "From Institution to Community: A New Process?" *Mental Retardation* 14 (1976), no. 6: 25-28.

DeVellis, R. F. "Learned Helplessness in Institutions." *Mental Retardation* 15 (1977), no. 5:10-13.

Farber, Bernard. "Effects of a Severely Mentally Retarded Child on Family Integration." *Society for Research in Child Development* 24 (1959), no. 2.

Golden, D., and Ho, C. C. *Behavior Modifications Inservice Training and Training Programs.* Orient, Ohio: Orient State Institute, 1973.

Grossman, H., ed. *Manual of Terminology and Classification in Mental Retardation.* Washington, D.C.: American Association on Mental Deficiency, 1973.

Houts, P. S., and Scott, R. A. *Goal Planning with Developmentally Disabled Persons.* Hershey, Pa.: Pennsylvania State University College of Medicine, 1975.

Keane, V. E. "The Incidence of Speech and Language Problems in the Mentally Retarded." *Mental Retardation* 10 (1972), no. 2: 3-5.

Keats, Sidney. *Cerebral Palsy.* Springfield, Ill.: Charles C. Thomas, 1970.

Lamb, Richard H. "Staff Burnout and Work with Long Term Patients." *Hospital and Community Psychiatry* 30 (1979), no. 6.

Leland, H., and Smith, D. E. *Mental Retardation: Present and Future Perspectives.* Worthington, Ohio: Charles A. Jones, 1974.

Luckey, R. E., and Shapiro, I. G. "Recreation: An Essential Aspect of Habilitative Programming." *Mental Retardation* 12 (1974), no. 5:33-35.

Maslach, Christina. "Burned Out." *Human Behavior Magazine,* 1976.

Munro, J. Dale. "Preventing Front Line Collapse in Institutional Settings." *Hospital and Community Psychiatry* 31 (1980), no. 3.

Murphy, L. B., and Leeper, E. M. *Language Is for Communication.* Washington, D.C.: U.S. Government Printing Office, 1974.

National Association for Retarded Children. *Residential Programming for Mentally Retarded Persons.* Vol. 3. *Developmental Programming in the Residential Facility.* Arlington, Tex.: 1972.

Nirje, B. "The Normalization Principle and Its Human Management Implications." In *Changing Patterns in Residential Services for the Mentally Retarded,* edited by R. B. Kugel and W. Wolfensberger. Washington, D.C.: President's Committee on Mental Retardation, 1969.

Parham, J. D.; Rude, L.; and Bernanke, P. *Individual Program Planning with Developmentally Disabled Persons.* Lubbock, Tex.: Texas Tech University, 1977.

Pines, Ayala, and Maslach, Christina. "Characteristics of Staff Burn Out in Mental Health Settings." *Hospital and Community Psychiatry* 29 (1978), no. 4.

Quilitch, H. R., and Grey, J. D. "Purposeful Activity for the PMR." *Mental Retardation* 12 (1974), no. 6:28-29.

Robinson, N. M., and Robinson, H. B. *The Mentally Retarded Child: A Psychological Approach.* New York: McGraw-Hill, 1976.

Roos, P. "Current Issues in Residential Care." Arlington, Texas: National Association of Retarded Children, 1969.

Sakel, Manfred. *Epilepsy.* New York: Philosophical Library, 1958.

Silverstein, Alvin, and Silverstein, Virginia B. *Epilepsy.* Philadelphia: J. B. Lippincott, 1975.

Sulzer-Azaroff, B., and Mayer, G. R. *Applied Behavioral Analysis Procedures with Children and Youth.* New York: Holt, Rinehart & Winston, 1977.

Throne, J. M. "Normalization through the Normalization Principle: Right End, Wong Means." *Mental Retardation* 13 (1975), no. 5:23-25.

United Cerebral Palsy Association. *Cerebral Palsy–Facts and Figures.* New York: 1979.

Vanderheiden, G., and Harris-Vanderheiden, D. "Communication Techniques and Aids for the Nonverbal Severely Handicapped." *Communication: Assessment and Intervention Strategies,* edited by L. Lloyd. Baltimore: University Park Press, 1976.

Watson, L. S. *How to Use Behavior Modification with Mentally Retarded and Autistic Children.* Columbus, Ohio: Behavior Modification Technology, 1972.

Wolfensberger, W. "The Origin and Nature of Our Institutional Models." In *Changing Patterns in Residential Services for the Mentally Retarded,* edited by R. B. Kugel and W. Wolfensberger. Washington, D.C.: President's Committee on Mental Retardation, 1969.

Wolfensberger, W., and Menolascino, F. J. In *Psychiatric Approaches to Mental Retardation,* edited by F. J. Menolascino, pp. 475–92. New York: Basic Books, 1970.

Wolfensberger, W.; Nirje, B.; Olshansky, Perske, S.; and Roos, P. *The Principle of Normalization in Human Services.* Toronto, Ontario, Canada: National Institute on Mental Retardation, 1972.

Wyne, M. D., and O'Connor, P. D. *Exceptional Children.* Lexington, Mass.: D. C. Heath, 1979.

Appendix A

4-1. a) This statement is not specific enough. We do not know what clothing John is to put on, how often, or when he is to do it. One alternative to this statement might be "John will put on his pants, shirt, socks, and shoes daily before going to breakfast."

b) This statement does not tell when, how, how long, or with whom John is to interact. An alternative might be "When given a ball, John will play catch with one other resident for five minutes."

c) This statement does not include how accurate we expect John to be, or how exact (to the hour or to the minute). One way of stating this objective more specifically would be "John will tell time to the quarter hour with 80 percent accuracy."

d) This statement describes what staff will be doing rather than what John will be doing. A way of rewriting this might be "John will go to the dining room on his own for each meal every day."

7-2. a) Washing face
1. Client will turn on the tap.
2. Client will pick up washcloth and put it in the water.
3. Client will pick up soap and rub it onto the cloth.
4. Client will put soap back in soapdish.
5. Client will wash lower part of face with the cloth.
6. Client will wash upper part of face with the cloth.

7. Client will put the cloth in the water, rinse out the soap, and wring out the excess water.

8. Client will rinse off the soap on the lower part of the face.

9. Client will rinse off the soap on the upper part of the face.

10. Client will put the cloth in the water, rinse out the soap, and wring out the excess water.

11. Client will return the washcloth to the proper place.

b) Putting on socks

1. Client will pull up sock if already over the heel.

2. Client will pull up sock if started on toes.

3. Client will pull up sock if handed in the correct position.

4. Client will put on sock unassisted.

c) Washing hands

1. Client will turn the water on.

2. Client will put both hands in the water.

3. Client will take hands out of the water and pick up soap.

4. Client will rub soap between hands until they are lathered.

5. Client will return soap to soapdish.

6. Client will rinse soap from hands.

7. Client will dry hands.

9-3. 1. The client will put one ring on the stick.

2. The client will put two rings on the stick when placed on table in order.

3. The client will put three rings on the stick when placed on table in order.

4. The client will put two rings on the stick when placed on table in random order (and can correct mistakes if necessary).

5. The client will put three rings on the stick when placed on table in random order (and can correct mistakes if necessary).

6. The client can put four rings on the stick when placed on table in random order (and correct mistakes if necessary).

7. The client can put five rings on the stick when placed on table in random order (and correct mistakes if necessary).

8. The client can put all the rings on the stick when placed on table in random order (and correct mistakes if necessary). (Golden and Ho 1973)

Appendix B

SOCIAL-SEXUAL EDUCATION TRAINING
MATERIALS AND GUIDES

Bass, M. *Developing Community Acceptance of Sex Education for the Mentally Retarded.* New York: SIECUS, 1972.

_____ , ed. *Sexual Rights and Responsibilites of the Mentally Retarded.* Proceedings of the Conference of the American Association on Mental Deficiency, Region IX. Newark, Del.: 1972.

Bidgood, F. "Sexuality and the Handicapped." *SIECUS Report – Special Issue on the Handicapped* (New York) 2:3, 1976.

De La Cruz, R., and LaVeck, G., eds. *Human Sexuality and the Mentally Retarded.* New York: Brunner/Mazel, 1973.

Edmonson, B.; McCombs, K.; and Wish, J. "What Retarded Adults Believe About Sex. *American Journal of Mental Deficiency* 84, 6, (1979, no. 1:11-18.)

Erikson, E. *Childhood and Society.* New York: Norton, 1950.

Fischer, H.; Krajicek, M.; and Borthick, M. *Sex Education for the Developmentally Disabled: Guide for Parents, Teachers, and Professionals.* Baltimore: University Park Press, 1974.

Gordon, S. *Facts About Sex.* New York: John Day, 1970.

_____ . Facts About V.D. for Today's Youth. New York: John Day, 1973.

_____ . *On Being the Parent of a Handicapped Youth.* New York: New York Association for Brain Injured Children, 1973.

_____ . *The Sexual Adolescent.* North Scituate, Mass.: Duxbury Press, 1973.

_____ . *Sexual Rights for People . . . Who Happen To Be Handicapped.* New York: Ed-U Press; Center on Human Policy, 1974.

_____. *Ten Heavy Facts About Sex.* New York: Ed-U Press, 1975.

Kempton, W. *A Guide to Sex Education for Persons with Disabilities That Hinder Learning.* Philadelphia: Planned Parenthood Association of Southeastern Pennsylvania, 1971.

_____. *A Teacher's Guide to Sex Education for Persons with Learning Disabilities.* North Scituate, Mass.: Duxbury Press, 1973.

_____. *Techniques for Leading Group Discussion on Human Sexuality.* Philadelphia: Planned Parenthood Association of Southeastern Pennsylvania, 1971.

Kempton, W.; Bass, M.; and Gordon, S. *Love, Sex, and Birth Control for the Mentally Retarded: A Guide for Parents.* Philadelphia: Planned Parenthood Association of Southeastern Pennsylvania and Family Planning and Population Center, 1971.

Meyers, R. *Like Normal People.* New York: McGraw-Hill, 1978.

Patullo, A. *Puberty in the Girl Who is Retarded.* New York: National Association for Retarded Children, 1969.

SIECUS. *Sex Education for the Handicapped.* New York: Human Science Press, n.d.

SIECUS and the American Alliance for Health, Physical Education, and Recreation (AAHPER). *A Resource Guide in Sex Education for the Mentally Retarded.* New York: Human Science Press, 1971.

Special Education Curriculum Development Center. *Social and Sexual Development.* Iowa: 1972.

Thaller, K., and Thaller, B. *Sexuality and the Mentally Retarded.* Washington, D.C.: Office of Economic Opportunity, 1973.

Wish, J.; McCombs, K.; and Edmonson, B. *The Socio-Sexual Knowledge and Attitudes Test.* Chicago: The Stoelting Company, 1979.

Wolfensberger, W. *Normalization.* Toronto: National Institute on Mental Retardation, 1972.

Index